ENVIRONMENTAL REMEDIATION TECHNOLOGIES, REGULATIONS AND SAFETY

DDT

PROPERTIES, USES AND TOXICITY

ENVIRONMENTAL REMEDIATION TECHNOLOGIES, REGULATIONS AND SAFETY

Additional books in this series can be found on Nova's website under the Series tab.

Additional e-books in this series can be found on Nova's website under the eBook tab.

ENVIRONMENTAL REMEDIATION TECHNOLOGIES, REGULATIONS AND SAFETY

DDT

PROPERTIES, USES AND TOXICITY

KATHLEEN SANDERS
EDITOR

NOVINKA
New York

NOTICE TO THE READER

Library of Congress Cataloging-in-Publication Data

ISBN: 978-1-53610-009-9

Published by Nova Science Publishers, Inc. † New York

CONTENTS

PREFACE

Dichloro-diphenyl-trichloroethane (DDT) is the world famous synthetic organochlorine broad-spectrum insecticide used in the control of mosquito vectors of malaria and insect pests of agriculture in developing countries, due to its low cost and high efficiency. This book discusses properties, uses and toxicity of DDT. Chapter One reports the empty-orbital electronic structure and experimental data of dissociative electron attachment (DEA) to the gas-phase molecules DDT and its principal metabolite DDE, which possesses good electron-withdrawing abilities. Chapter Two reviews the past, present and future of DDT and pyrethroid resistance in *Aedes aegypti* (L.) and *Aedes albopictus* (Skuse). Chapter Three studies the photocatalytic selective oxidation of DDT.

Chapter 1 – This Chapter reports the empty-orbital electronic structure and experimental data of dissociative electron attachment (DEA) to the gas-phase molecules DDT (1,1'-(2,2,2-trichloroethane-1,1-diyl)bis(4-chlorobenzene)) and its principal metabolite DDE (1,1-bis-(4-chlorophenyl)-2,2-dichloroethene), which possesses good electron-withdrawing abilities. Electron transmission spectroscopy (ETS) is employed to study formation of temporary negative ions (TNIs) by electron addition to vacant molecular orbitals of neutral molecules. Fragments formed by dissociation of TNIs, in kinetic competition with extra electron detachment, were detected by means of DEA spectroscopy. The experimental findings are supported by density functional theory (DFT) calculations. According to ideas put forward by James Lovelock in the early sixties of the XX century, these findings play a vital role to describe toxic effects produced *in vivo* by free radical formation followed by deactivation of enzymes by chemicals with high electron affinity, as in the case of the model toxicant carbon tetrachloride (CCl_4). The present DEA results indicate that, in close analogy with CCl_4, a chlorine negative ion is effectively eliminated from DDT “activated” by capture of an extra electron. Provided that a similar process can take place in the cellular ambient under conditions of excess negative charge, this would give rise to formation of DDE and the second most abundant metabolite DDD (1-chloro-4-[2,2-dichloro-1-(4-chlorophenyl)ethyl]benzene) through generation of intermediate dechlorinated radicals. Two possible mechanisms for attachment of an extra electron to DDT in cells are discussed. In the first one, where e^- derives from the mitochondrial electron transport chain, DDD formation is associated with hydrogen atom abstraction from neighboring lipids by $[DDT – Cl]^\bullet$ neutral radicals, eventually initiating chain reactions of peroxidation of biomembranes. In the second one, e^- can originate from the catalytic cycle of cytochrome P450, where dechlorinated radicals can be tightly bound to an active center causing “suicide deactivation” of P450 enzymes.

Chapter 2 – *Aedes aegypti*, an important vector of viral diseases such as yellow fever, dengue, chikungunya, and zika, is thought to have originated in the African continent. The subspecies that has been domesticated and adapted to anthropogenic environments (*Aedes aegypti aegypti*) expanded its habitat around human domiciles and has been dispersed by human movement. Worldwide eradication programs with organochlorine insecticides in 1940s and 1950s may have caused high selection pressure on mosquitoes and expanded the distribution of insecticide-resistant *Ae. aegypti* populations. DDT resistance in *Ae. aegypti* was first reported in Caribbean countries in the 1950s, and that resistance persists in almost all regions that had achieved *Ae. aegypti* eradication, despite the fact that DDT is no longer used. *Aedes albopictus* inhabits tropical regions but is also found in temperate areas, including Japan, Europe, America, and Australia. Although dengue hemorrhagic fever has not been prevalent in Japan for the past 50 years and a chikungunya outbreak has never been reported in Japan, the possibility of viral disease outbreaks exists because of the widespread distribution of the vector *Ae. albopictus*. DDT resistance in this species seems to be common in Japan, while resistance to other insecticides such as pyrethroids, organophosphates, and carbamates is uncommon. It is noteworthy that there has been a worldwide focus for the past three decades on the dispersal of shipping containers of used tires that carry breeding mosquitoes. The mosquito species that has played the leading part in these events is *Ae. albopictus*, which together with four other Japanese mosquito species is thought to have arrived at the western coastal ports of the United States in used tires by 1983, becoming widely distributed in the United States and Brazil by 1986. Since 1986, tire shipments infested with *Ae.*

albopictus have been found in South and Central America, South and West Africa, Oceania, and Europe. Used tire trading occurs worldwide, with complicated commercial networks in many countries including South Africa, Kenya, Uganda, Niger, Nigeria, and Ghana. The history of the worldwide invasion of *Ae. aegypti* in association with human activity may be longer than that of *Ae. albopictus*. The worldwide transportation of vector mosquitoes in association with human activity may result in the introgression of DDT and other insecticide-resistance genes. DDT has played an important role as the most effective insecticide for indoor residual spraying (IRS). The Stockholm Convention seeks the elimination of DDT. There is a concern, however, that a ban on DDT use could adversely affect the malaria burden, since many countries have relied on the effectiveness and low cost of DDT treatment. The use of DDT is now recommended only where the intervention is appropriate and effective in the local epidemiologic situation. The continued efficacy of DDT, on the other hand, seems to have given rise to DDT resistance and cross-resistance to pyrethroids in *Ae. aegypti* and *Ae. albopictus*, as well as in *Anopheles* mosquitoes.

Chapter 3 – DDT (dichlorodiphenyltrichloroethane), used as a pesticide, is a persistent pollutant and rests one of the main dangers to human health and to the environment. Being chemically stable in ordinary conditions, its oxidation or oxidative degradation, especially, in large amounts or in industrial scale, is problematic in certain aspects. In this work it has been presented the experimental results of the direct oxidation of DDT with dioxygen, cheapest oxidant agent, using as a heterogeneous photocatalyst the fully characterized metal organic complex dioxo-Mo(VI)-dichloro[4,4′-dicarboxylato-2,2′-bipyridine], anchored on the surface of a TiO_2 by chemical bonds. The oxidation of DDT was carried out as a two stage cyclic process, involving: (i) the direct interaction of Mo-complex/TiO_2 with DDT under UV-irradiation ($\lambda = 253.7$ nm), and, (ii) regeneration of the photocatalyst with dioxygen in the dark conditions. These two stages of the process were separated temporally. It has been shown that the dioxo-Mo complex, anchored on the TiO_2, plays the role of photocatalyst in selective oxidation of DDT to dicofol (2,2,2-trichloro-1,1-*bis*-(4-chlorophenyl)ethanol) in chosen "mild" conditions. About 35% of the initial DDT was oxidized and the yield of dicofol was 21 mol %, during in 5 consecutive experimental cycles (32 h). It has been shown also that the formation of products, other than dicofol, is intermediated mainly by the oxidative and partially reductive decomposition of dicofol. They are dechlorinated (28-30%) and non-dechlorinated or partially dechlorinated organic compounds (34-36%), as well as the products of the complete degradation to CO_2 H_2O and HCl (15-17%). On the basis of kinetic data and analysis of products it has been proposed the probable mechanism of the photocatalyzed oxidation of DDT in the mentioned conditions. The scheme of the reaction mechanism includes the primary reaction of the oxo-atom transfer from the Mo-complex to tertiary carbon-atom of DDT by the formation of dicofol. About twofold increase of the turnover number of the reaction (Δ[DDT]/[anchored complex]) in second and further catalytic cycles evidences that the re-oxidation with dioxygen of Mo(IV) metal centers, formed in the first stage of catalytic cycle, to Mo(VI) may occur by the formation of intermediate Mo(VI)-oxo-peroxo moieties of the anchored complex.

In: DDT
Editor: Kathleen Sanders
ISBN: 978-1-53610-009-9

Chapter 1

Electron Acceptor Properties of DDT and Molecular Mechanism of Its Toxicity

***Stanislav A. Pshenichnyuk*[1,2,*], *Alberto Modelli*[3,4] *and Alexei S. Komolov*[2]**

[1]Institute of Molecule and Crystal Physics, Ufa Research Centre, Russian Academy of Sciences, Ufa, Russia
[2]Physics Faculty, St. Petersburg State University, St. Petersburg, Russia
[3]Università di Bologna, Dipartimento di Chimica "G. Ciamician," Bologna, Italy
[4]Centro Interdipartimentale di Ricerca in Scienze Ambientali, Ravenna, Italy

Abstract

This Chapter reports the empty-orbital electronic structure and experimental data of dissociative electron attachment (DEA) to the gas-phase molecules DDT (1,1'-(2,2,2-trichloroethane-1,1-diyl)bis(4-chlorobenzene)) and its principal metabolite DDE (1,1-bis-(4-chlorophenyl)-2,2-dichloroethene), which possesses good electron-withdrawing abilities. Electron transmission spectroscopy (ETS) is employed to study formation of temporary negative ions (TNIs) by

* Corresponding author: E-mail: sapsh@anrb.ru.

electron addition to vacant molecular orbitals of neutral molecules. Fragments formed by dissociation of TNIs, in kinetic competition with extra electron detachment, were detected by means of DEA spectroscopy. The experimental findings are supported by density functional theory (DFT) calculations. According to ideas put forward by James Lovelock in the early sixties of the XX century, these findings play a vital role to describe toxic effects produced *in vivo* by free radical formation followed by deactivation of enzymes by chemicals with high electron affinity, as in the case of the model toxicant carbon tetrachloride (CCl_4).

The present DEA results indicate that, in close analogy with CCl_4, a chlorine negative ion is effectively eliminated from DDT "activated" by capture of an extra electron. Provided that a similar process can take place in the cellular ambient under conditions of excess negative charge, this would give rise to formation of DDE and the second most abundant metabolite DDD (1-chloro-4-[2,2-dichloro-1-(4-chlorophenyl) ethyl]benzene) through generation of intermediate dechlorinated radicals. Two possible mechanisms for attachment of an extra electron to DDT in cells are discussed. In the first one, where e^- derives from the mitochondrial electron transport chain, DDD formation is associated with hydrogen atom abstraction from neighboring lipids by $[DDT - Cl]^\bullet$ neutral radicals, eventually initiating chain reactions of peroxidation of biomembranes. In the second one, e^- can originate from the catalytic cycle of cytochrome P450, where dechlorinated radicals can be tightly bound to an active center causing "suicide deactivation" of P450 enzymes.

Keywords: negative ion states, resonances, dissociative electron attachment, toxicity, neurotoxicity, free radical mechanism, mitochondrial electron transfer, electron leak, mitochondrial disorder, biomembrane disruption, enzyme deactivation

INTRODUCTION

1,1,1-trichloro-2,2-bis(4-chlorophenyl)ethane (DDT) is the world famous synthetic organochlorine broad-spectrum insecticide used in the control of mosquito vectors of malaria and insect pests of agriculture in developing countries, due to its low cost and high efficiency [1, 2]. Its action, which is viewed as one of the most interesting phenomena in biochemical mechanisms of pesticide activity, has been widely discussed [3, 4]. In spite of its widespread usage since the discovery of its insecticidal action in 1939 by the Swiss chemist Paul Hermann Müller, DDT was found to be dangerous for the wildlife, being toxic to birds, fishes, amphibians and aquatic microorganisms. Moreover, owing to its stability, it can be accumulated in ecosystems [5, 6]. DDT is suspected to be carcinogenic, influences female reproductive organs, causes endocrine disruption and long-term adverse effects in human organism [6, 7] where it tends to accumulate in fatty tissues [8]. DDT can permeate all forms of life (humans, plants, animals), and environment (water, air, and soil) traveling through the food chain, so that its agricultural use was reasonably banned in 1972 in the United States, and a worldwide ban was then formalized under the Stockholm Convention on Persistent Organic Pollutants [9]. These regulations caused a decrease of the levels of DDT bioaccumulation in humans. However, the long-term adverse effects of dormant residues present in most living species all over the world are still a topical and debated scientific issue.

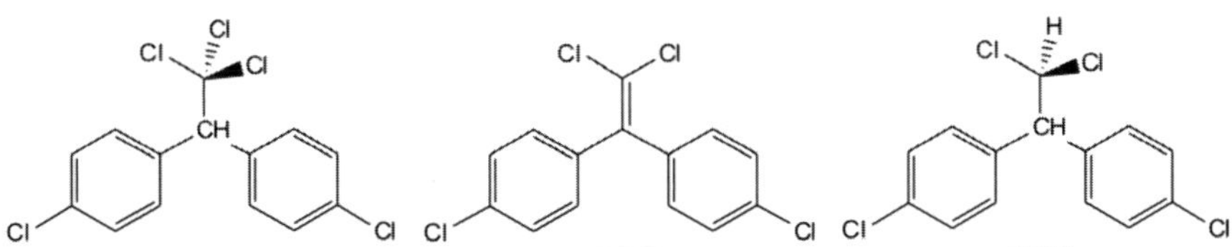

Figure 1. Structures of DDT and its main metabolites DDE and DDD.

DDT is metabolized primarily in liver via reductive dehydrochlorination and dechlorination (the latter being the first and rate-limiting step in its metabolism) [10] with production of 1,1-dichloro-2,2-bis(4-chlorophenyl)ethene (DDE) and 1,1-dichloro-2,2-bis(4-chlorophenyl)ethane (DDD, see structural formulae in Figure 1), respectively, as main metabolites, as firstly discovered in yeast [11] and later in rats [12]. Cytochrome P450 enzymes were found to be responsible for reductive dechlorination of DDT and its structural analogs in liver microsomes of rats in vitro. However, nonenzymatic metabolism was also suggested with involvement of reduced flavin (see ref. 13 and references therein). The exact molecular mechanisms of DDT dechlorination is unknown [14]. Due to their stability and capacity to be accumulated in fat, DDT primary metabolites are also largely present in living organisms all over the world [8].

Although the molecular mechanisms which cause toxic effects are unknown, neurophysiological and reproductive dysfunctions are attributed to DDT, and the early effect of carcinogenicity is ascribed to the interaction with plasma membrane and alteration of its structure and functions [15]. DDT and DDE were found to damage the structure of mitochondrial membranes [16] and reveal an unwanted and harmful activity in the intermembrane space, both being potent inhibitors of the mitochondrial electron transport chain [17]. Additionally, DDT blocks succinate dehydrogenase (the respiratory Complex II) and electron transfer between cytochromes b and c, leading in turn to a reduction of ATP synthesis [18]. DDT can also inhibit gap-junction intercellular communications [19, 20], thus stimulating tumor promotion [21]. Some evidence was found that irreversible DNA damage (genotoxicity) can be caused by DDT, the effect being about one half of that produced by DDE [22].

Although DDT acts as a nerve poison in both insects and mammalians, causing hyperexcitability and tremors of increasing severity as main symptoms [1], some of its structural and electronic properties are similar to those of the model toxicant carbon tetrachloride (CCl_4), whose action is mainly associated with hepatotoxicity, namely, fatty degeneration of the liver and centrilobular necrosis [23]. The corresponding liver cells injury is supposed to be mediated by one-electron reductive dehalogenation of CCl_4 (electron capture by CCl_4, followed by dissociation of a chloride fragment [23]) to trichloromethyl free radical ($CCl_3^•$), catalyzed by cytochrome P450 in endoplasmic reticulum, some secondary mechanisms being responsible for plasma membrane disruption [23-25]. Some other studies report on similar dehalogenation mechanisms driven by cytochrome P450 in the liver microsomal fraction of mammalian species, as observed with a series of halogenated compounds [26-28].

Therefore, the well-known one-electron reductive dehalogenation mechanism of carbon tetrachloride (frequently used as a model for toxic activity) by the P450 monooxygenase system is the first step leading to hepatotoxicity [23, 29], and infers a connection between toxicity and dissociative electron attachment (DEA), i.e., the process $CCl_4 + e^- \rightarrow CCl_4^{(*)-} \rightarrow CCl_3^• + Cl^-$, where $CCl_4^{(*)-}$ stands for temporary (usually excited) negative ion (TNI). This mechanism of toxicity has also been reported earlier by Gregory [30], in line with more recent findings [31-35]. Reductive dehalogenation of CCl_4 by cytochrome P450 generates trichloromethyl free radicals, which initiate the toxic process [25, 31, 32]. Then, irreversible cell damage is driven by stimulation of lipid peroxidation and binding of electrophilic radicals to membrane lipids [33]. Incubation with CCl_4 results in a decrease (almost 50%) of the content of unsaturated fatty acids in microsomal lipids and exerts a pronounced inhibitory effect on the exocytosis of macromolecules [34]. This mechanism likely proceeds through hydrogen abstraction from neighboring lipids by the highly reactive $CCl_3^{\bullet}$ radical [30], leading to a breakdown of the membrane lipids and, finally, liver damage [35].

The above mentioned free radical mechanism of injury of living tissues is in line with ideas formerly formulated by Lovelock [36-38] on a connection between electron affinity and biological activity of electron-withdrawing molecules, as well as his conclusion that the toxicity of these compounds is due to their ability to irreversibly trap electrons, with production of potentially dangerous species via DEA [36]. Therefore, the knowledge of DEA properties (a fundamental process also viewed as one of the simplest chemical reactions) for a specific molecular target is considered to be of great importance to interpret its bioactivity. Indeed, gas-phase DEA is the counterpart of dissociative electron transfer (DET) in solution [39-41], plausibly expected to occur near electron transport pathways in the cells as the first step towards a multifaceted biochemistry of various classes of active species including ecotoxicants [42], anti-inflammatory drugs [43], pesticides [44], flavonoids [45], spinochromes [46], plant hormones [47], and polyphenolic compounds [48, 49].

DEA [50-53] can be investigated in detail in vacuo by means of the complementary experimental techniques electron transmission spectroscopy (ETS) [54] and DEA spectroscopy [55] with the support of suitable quantum-chemical calculations. Such a complex approach allows to reveal the structure of negative and neutral fragments and the energies of their formation, i.e., the energy positions of electron-molecule resonances [50], which are related to the energies of the empty molecular orbitals (MOs) of the target molecule. A solid matter can also be subjected to low-energy electron transmission studies using the total current spectroscopy (TCS) technique [56]. The TCS results have been shown quite supportive for the analysis of the empty energy bands corresponding to the empty MO in organic molecular solids [56-59]. When electron attachment and DEA processes take place in vivo in a cytosolic liquid phase, the solvent effects on the empty-level structure are of course to be accounted for. Christophorou and Hadjiantoniou [60] suggested that a shift of the resonance positions to lower energies by nearly 1 eV with respect to the isolated molecules is often an approximate correction, and concluded that only anion states lying below 1 eV in the gas-phase would participate to DEA reactions in cellular tissues. When electron attachment in cellular tissues occurs at interphase boundaries (lipids-ferments-cytosol) it involves weakly bound quasi-free electrons [61] whose binding energy approximates the "electron attachment window" for DEA to many polyatomic molecules possessing stable negative ion states [61, 62]. Provided that cellular electrons are acquired from enzymatic active centers, there is a close relation between DET processes and gas-phase DEA, because the binding pockets of proteins are usually hydrophobic and contain a very small number of water molecules [63].

The present Chapter reports on structures of species produced by DEA to DDT and DDE under gas-phase conditions. Density functional theory (DFT) calculations are employed to assign the observed resonant states to the corresponding virtual MOs of the isolated molecules. In order to discuss possible correlations with cellular processes, the calculations were also performed simulating the solvent (water) effect in the framework of the polarizable continuum model (PCM), thus modeling the conditions of living tissues. The fragments produced in the DEA processes by DDT and DDE were classified as (i) potentially dangerous radicals able to abstract a hydrogen atom from the neighboring lipids, (ii) molecular and anionic species whose fate in cellular tissues is not obvious but could initiate a cascade of unexpected biochemical reactions. Likely after-effects of generation of these species in living tissues, such as (i) lipid membrane peroxidation and rupture, (ii) mitochondrial DNA mutations through formation of DNA-adducts, (iii) deactivation of enzymes, are discussed. In particular, it should be noted that the products of DEA to xenobiotic species can likely weaken the native defense system, not trained for deactivation of these unexpected active agents. The present investigation can thus give some more insight into the mechanisms of toxic action.

EXPERIMENTAL AND COMPUTATIONAL METHODS

Electron transmission spectroscopy. ETS is one of the most suitable means for detecting the formation of TNI states, as a function of the incident electron energy, which appear as sharp structures in the electron-molecule scattering cross section [50, 54, 55]. Because electron attachment is rapid with respect to nuclear motion, TNIs are formed with the equilibrium geometry of the neutral molecule. The energies at which electron attachment occurs are properly denoted as vertical attachment energies (VAEs) and are the negatives of the vertical electron affinities (EA_vs).

Our electron transmission apparatus is in the format devised by Sanche and Schulz [54], and has been previously described [64]. Briefly, a magnetically collimated electron beam is passed through a collision cell containing the target gas. The first derivative of the electron current transmitted through the gas sample is measured with a synchronous lock-in amplifier. Each resonance, associated with formation of a temporary negative ion, is characterized by a minimum and a maximum in the derivative signal. The energy of the midpoint between these features is assigned to the VAE. The electron beam resolution was about 50 meV. The energy scale was calibrated with reference to the $(1s^12s^2)$ ^2S anion state of He. The estimated accuracy of the measured VAEs is ±0.05 or ±0.1 eV, depending on the number of decimal digits reported.

Dissociative electron attachment spectroscopy. Additional information on the fate of molecular TNIs can be supplied by DEA spectroscopy which measures the yield of mass-selected negative ions as a function of the impact electron energy. When suitable energetic conditions apply, the decay of unstable molecular anions formed by resonance can follow a dissociative channel which generates a negative fragment and a neutral radical, in kinetic competition with simple re-emission of the extra electron.

A general overview of DEA spectroscopy may be found elsewhere [51, 53, 55]. Our magnetic mass spectrometer was described in detail previously including a schematic representation and description of specific conditions [65]. Briefly, a magnetically collimated electron beam of defined energy was passed through a collision cell containing a vapor of the substance under investigation, under single-collision conditions. A current of magnetically mass-selected negative ions was recorded as a function of the incident electron energy in the 0-14 eV energy range. The electron energy scale was calibrated with the SF_6^- signal at zero energy, generated by attachment of thermal electrons to SF_6. The full width at half maximum (fwhm) of the electron energy distribution was 0.4 eV, and the accuracy of the measured peak positions was estimated to be ±0.1 eV.

A procedure for experimental evaluation of the electron detachment time (τ_a) from mass-selected long-lived negative ions, i.e., the process referred to as autodetachment, was previously proposed [66] for time-of-flight experiments. A detailed description of a similar method suitably modified for magnetic mass filters may be found elsewhere [53, 65]. As highlighted earlier [67, 68], the evaluated τ_a values are averaged over the vibrational states of the target molecule, convoluted with the electron energy distribution, and are also significantly affected by the accessible experimental “time window,” that is, the time interval during which neutral particles formed by electron detachment can be detected. The dependence on the time window is due to the fact that the analyzed anion beam actually consists of anions with different electron detachment times, due to the energy spread of the incident electron beam and the different vibrational states of the target molecules. The measured τ_a is thus averaged over the anions detectable in a given time window, and of course a change in the latter affects the evaluated τ_a value.

The compounds under investigation are commercially available (Sigma-Aldrich # 31041 and 35487 for DDT and DDE, respectively) and were used without further purification (their notified purity is 99.7% and 99.9%, respectively). They were evaporated at 60-80 ^{0}C keeping the walls of the collision cell 10 ^{0}C hotter to prevent condensation.

Quantum-chemical calculations. DFT calculations were performed with the Gaussian 09 set of programs [69]. Evaluation of the virtual orbital energies (VOEs) of the neutral molecule was performed using the B3LYP hybrid functional [70] with the standard 6-31G(d) basis set. The EA_v was calculated as the difference between the total energy of the neutral and the lowest anion state, both in the optimized geometry of the neutral state, using the standard 6-31+G(d) basis set (which includes the minimum addition of diffuse functions). The adiabatic electron affinity (EA_a) was obtained as the energy difference between the neutral and the lowest anion state, each in its optimized geometry.

Regardless of the peculiar difficulties encountered for the description of anionic states [71], it has been demonstrated [72-74] that good linear correlations can be obtained between the VAEs measured in ETS and the corresponding VOEs of the neutral molecules calculated with basis sets which do not include diffuse functions. The scaling parameters are different for σ* and π* MOs. Moreover, it should be expected that a more accurate correlation would result if "training" compounds were employed that were structurally closer to the subject molecule. In the present study the linear equation ($VAE = 0.8065 \times VOE + 0.9194$) derived for the π* MOs of alternating phenyl and ethynyl groups [75] was employed to scale the π* VOEs. The scaling procedure was also successfully used for the peak assignment of the experimentally obtained density of the unoccupied states in small conjugated molecular materials [58, 59, 76]. The σ* VOEs were scaled using the equation ($VAE = 0.8111 \times VOE + 1.6097$) found for $\sigma^*_{C\text{-}Cl}$ orbitals in a series of mono- and polychloroalkanes [77]. It should be noted that the σ* VAEs evaluated with this procedure are generally less reliable than the π* VAEs, due to the smaller number of experimental data available for calibration of the scaling equations. The effects of a solvated environment were evaluated with the PCM supplied by the Gaussian 09 package.

RESULTS

Temporary negative ion states and geometries of neutral molecules. The electron transmission (ET) spectrum of DDT is reported in Figure 2, where the vertical bars locate the measured VAEs, i.e., the most probable incident electron energies at which anion formation occurs. At the maximum available temperature (110°C) of the collision chamber attainable with our ETS technique, the vapor pressure of DDE was not sufficient to give a proper attenuation of the electron beam, therefore the VAEs of DDE are derived only from the calculations (see Table 1). In order to assign the resonances observed in the ET spectrum of DDT to the corresponding empty MOs, theoretical calculations on the neutral state molecules were carried out at the B3LYP/6-31G(d) level of theory. Table 1 reports the VOEs of the lower-lying MOs of DDT and DDE, together with the VOEs scaled with empirical linear equations (SVOEs). The optimized geometries of the neutral ground states molecules of DDT and DDE are shown in Figure 3. Schematic representations of the lowest empty MOs of DDT and DDE, as supplied by B3LYP/6-31G(d), calculations are presented in Figure 4.

The DDT neutral molecule possesses six π^* MOs associated with the benzene rings. The first resonance displayed by the ET spectrum of DDT is located at 0.5 eV. The scaled VOEs predict the two lowest π^* anions to lie at 0.20 eV and 0.31 eV, and the lowest σ^* anion state at 0.41 eV. As mentioned above, the σ^* scaling is expected to be less reliable than the π^* scaling. In fact, according to the B3LYP/6-31+G(d) calculations the first anion state is stable, possesses $\sigma^*_{C\text{-}Cl}$ character and is mainly localized on the CCl_3 group. The first three anion states are thus plausibly not observed in the ET spectrum because they are stable or masked by the high-energy wing of the intense electron beam signal. According to the scaled VOEs the next two π^* anion states lie close to each other at 0.47 and 0.50 eV, and are thus associated with the resonance observed at 0.5 eV. Then the scaled VOEs locate four anion states with mainly $\sigma^*_{C\text{-}Cl}$ character in the 1.5-2.1 eV energy range (see Table 1). In agreement, the ET spectrum displays a broad (1.2 eV fwhm) and unresolved signal centered at 1.8 eV. The two higher-lying π^* anions give rise to the unresolved resonance centered at 4.0 eV.

The volatility of DDE was not sufficient to record a spectrum with our ETS apparatus. The present description of its empty level structure therefore relies on the scaled VOEs presented in Table 1. The first anion state, with mainly $\pi^*_{C=C}$ character, is predicted to be stable by 0.29 eV, in good agreement with the EA_v (about 0.4 eV) estimated as the neutral/anion total energy difference with B3LYP/6-31+G(d) calculations, as reported in Table 2. The next three π* scaled VOEs are relatively close to each other in the 0.1-0.5 eV energy range. Then the scaled VOEs predict a π* anion state at 1.35 eV, with antibonding character between the $\pi^*_{C=C}$ group and the adjacent benzene rings, followed by four $\sigma^*_{C\text{-}Cl}$ anion states in the 1.6-2.5 eV energy range. Finally, the two benzene higher-lying π* anions should lie at 3.7-3.8 eV.

Table 1. B3LYP/6-31G(d) VOEs, scaled values (SVOEs), and measured VAEs in DDT and DDE. All values are in eV

orbital		VOE		SVOE		VAE
DDT						
π_6^*		3.816		4.00		4.0
π_5^*		3.217		3.51		
$\sigma_5^*{}_{C\text{-}Cl}$		0.654		2.14		
$\sigma_4^*{}_{C\text{-}Cl}$		0.599		2.10		1.8
$\sigma_3^*{}_{C\text{-}Cl_3}$		0.033		1.64		
$\sigma_2^*{}_{C\text{-}Cl_3}$		-0.194		1.45		
π_4^*		-0.516		0.50		
π_3^*		-0.555		0.47		0.5
π_2^*		-0.758		0.31		
π_1^*		-0.891		0.20		
$\sigma_1^*{}_{C\text{-}Cl_3}$		-1.476		0.41		
DDE						
π_7^*		3.605		3.83		
π_6^*		3.417		3.68		
$\sigma_4^*{}_{C\text{-}Cl_2}$		1.059		2.47		
$\sigma_3^*{}_{C\text{-}Cl}$		0.625		2.12		
$\sigma_2^*{}_{C\text{-}Cl}$		0.609		2.10		
π_5^*		0.530		1.35		
$\sigma_1^*{}_{C\text{-}Cl_2}$		0.007		1.62		
π_4^*		-0.509		0.51		
π_3^*		-0.754		0.31		
π_2^*		-0.982		0.13		
$\pi_1^*{}_{C=C}$		-1.505		-0.29		

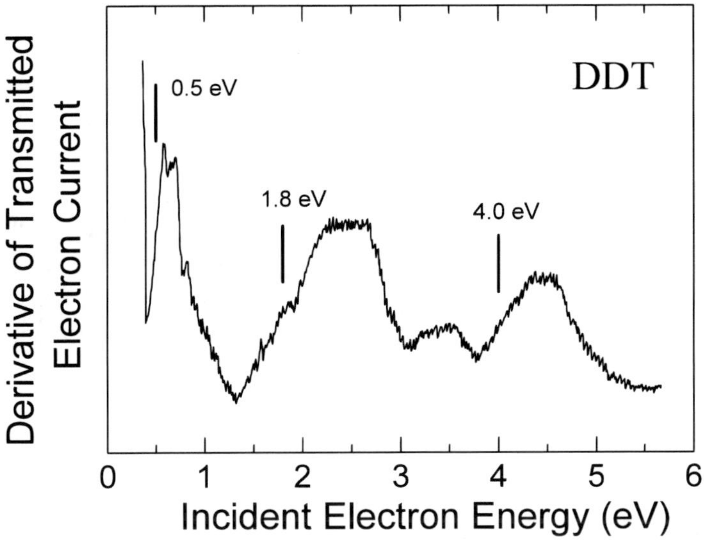

Figure 2. Electron transmission spectrum of DDT.

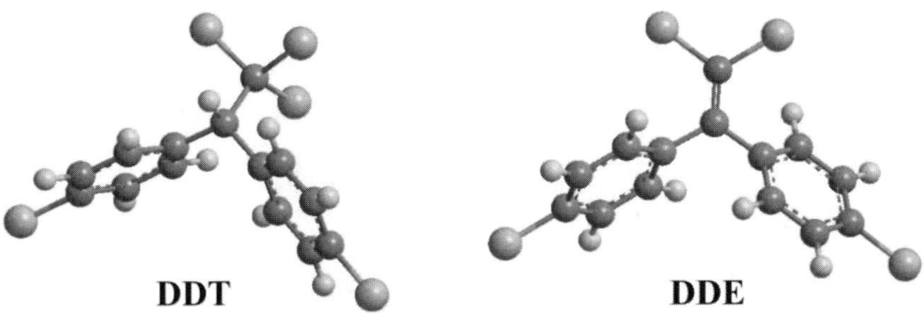

Figure 3. B3LYP/6-31+G(d) optimized geometries of the neutral ground states of DDT and DDE.

Fragmentation by low-energy electrons. Figure 5 reports the DEA spectra of the compounds under investigation, in the 0-14 eV incident energy range. The absence of distinct signals at incident electron energies >1.5 eV in both DDT and DDE appears at a glance as a rather unusual feature, so that essentially only electron attachment to the lowest-lying MOs contributes to fragmentation of the temporary molecular anions. In other words, attachment of electrons with epithermal energies essentially does not cause dissociation of the DDT and DDE molecules. In this regard, it can be recalled that zero-energy anion currents are associated with (vibrationally excited) stable anion states. Therefore, electron addition to the lowest four π* MOs and three σ* MOs of DDT and the lowest five π* MOs and first σ* MO of DDE can be associated with production of the anion fragments observed in their DEA spectra. Due to the involvement of several close-in-energy anion states and a relatively small electron energy resolution of our DEA measurements, the anion currents recorded as a function of the incident electron energy give rise to bands where possible different contributions are not resolved. Likely structures of fragment anions, and their neutral counterparts, formed by DEA to DDT and DDE are listed in Table 2. Their assignment is supported by the calculated thermodynamic energy thresholds in vacuo.

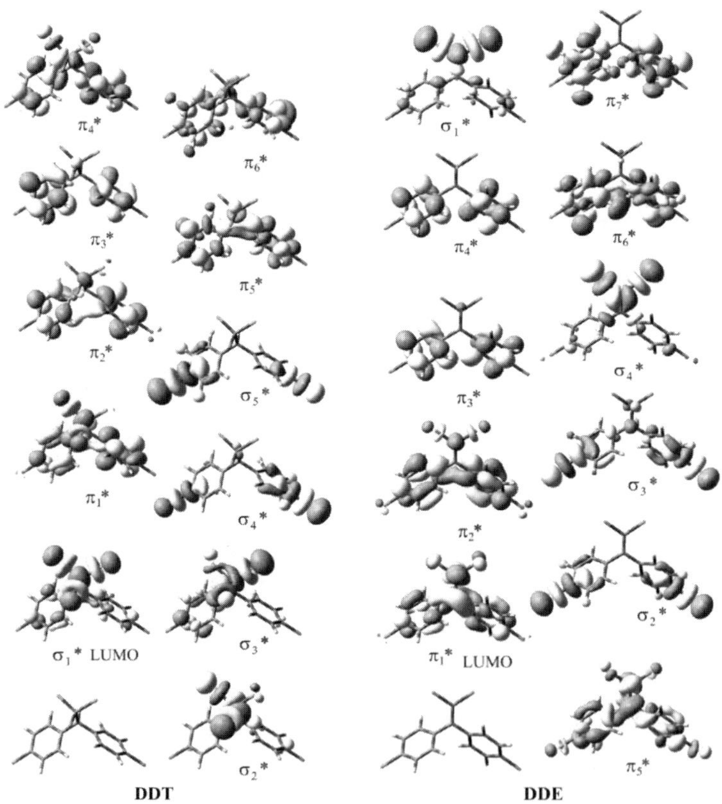

Figure 4. B3LYP/6-31G(d) localization properties of the lowest vacant MOs of neutral DDT and DDE.

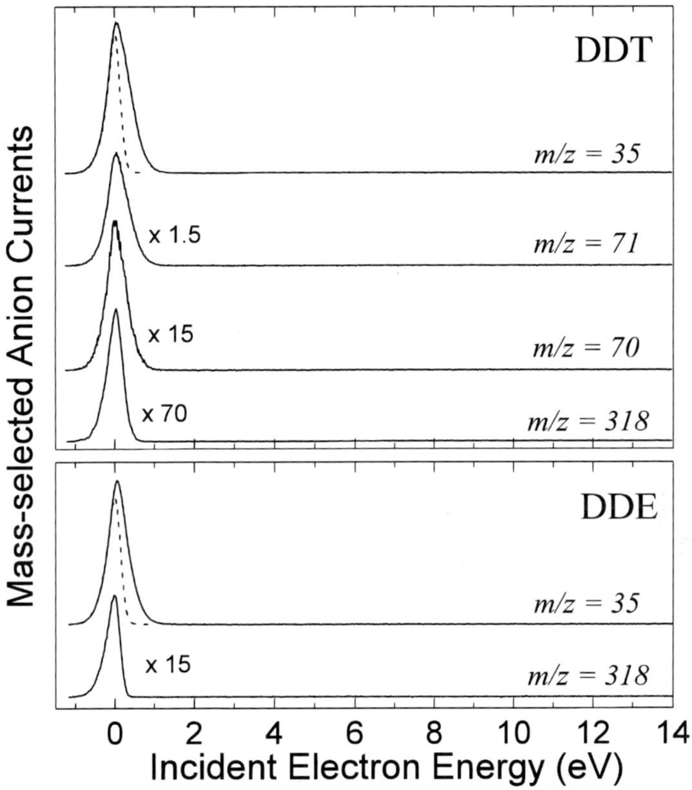

Figure 5. DEA spectra of DDT and DDE. The dashed line represents the current of SF_6^- formed by DEA to sulphur hexafluoride.

The dissociative pathways of the DDT TNIs were found to be simple and involve mostly the CCl_3 group. All the observed anion currents peak at zero energy (see Figure 5). The most abundant negative species is Cl^-, obtained by simple cleavage of a C–Cl bond. Elimination from the CCl_3 group is calculated to be energetically favored as compared with that from the aromatic ring (see Table 2). Only three more dissociative decay channels were detected in the DEA spectra. Two negative fragments with m/z = 71 and 70 are ascribed, respectively, to formation of Cl_2H^- and Cl_2^-. According to the calculations, the two channels lead to products, respectively, slightly and strongly more stable than DDT. Finally, a current at m/z = 318 is also observed, although with smaller intensity. In agreement, elimination of a neutral hydrogen chloride molecule to give an anion with the structure of DDE⁻ is calculated to be energetically favored (see Table 2).

Table 2. B3LYP/6-31+G(d) total electronic energies of fragments formed by DEA to DDT and DDE relative to that of the neutral ground states in vacuum and in water. The values in parentheses are corrected for zero-point vibrational energies

m/z	fragments	energy (eV)	
		vacuum	water
DDT			
354	DDT^- vert.	-0.41 (-0.55)	-1.90 (-2.05)
354	DDT^- adiab.	-1.97 (-2.05)	-4.07 (-4.15)
318	$[DDT - HCl]^- + HCl$	-0.73 (-1.03)	-3.16 (-3.43)
71	$Cl_2H^- + [DDT - Cl_2H]^\bullet$	-0.08 (-0.34)	-3.15 (-3.42)
70	$Cl_2^- + [DDT - Cl_2]$	-2.17 (-2.27)	-4.62 (-4.72)
35	$Cl^- + [DDT - Cl]^\bullet$	-1.10 (-1.18)	-4.07 (-4.15)
35	Cl^- (ring) + $[DDT - Cl]^\bullet$	0.23 (0.14)	-2.75 (-2.85)
DDE			
318	DDE^- vert.	-0.37 (-0.52)	-1.78 (-1.94)
318	DDE^- adiab.	-0.87 (-0.99)	-3.22 (-3.30)
35	$Cl^- + [DDE - Cl]^\bullet$	-0.27 (-0.35)	-3.24 (-3.32)
35	Cl^- (ring) + $[DDE - Cl]^\bullet$	0.23 (0.13)	-2.76 (-2.85)

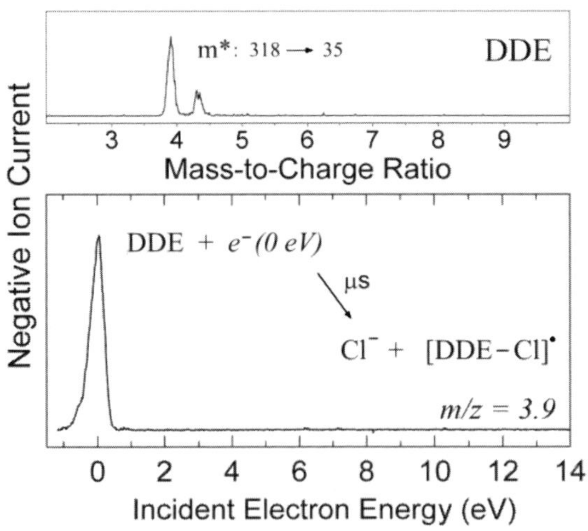

Figure 6. Experimental observation of a metastable (microseconds timescale) decay of the DDE molecular anion at zero energy, with elimination of a chloride anion.

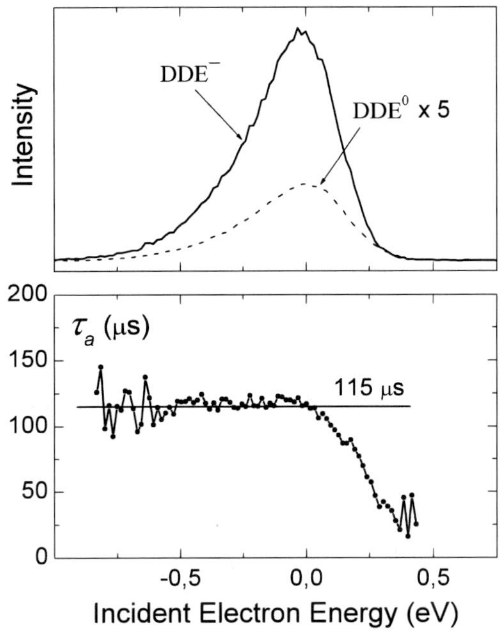

Figure 7. Current of DDE^{-}, signal of the corresponding neutral counterpart DDE^{0}, and evaluated mean electron detachment time τ_a from DDE^{-}.

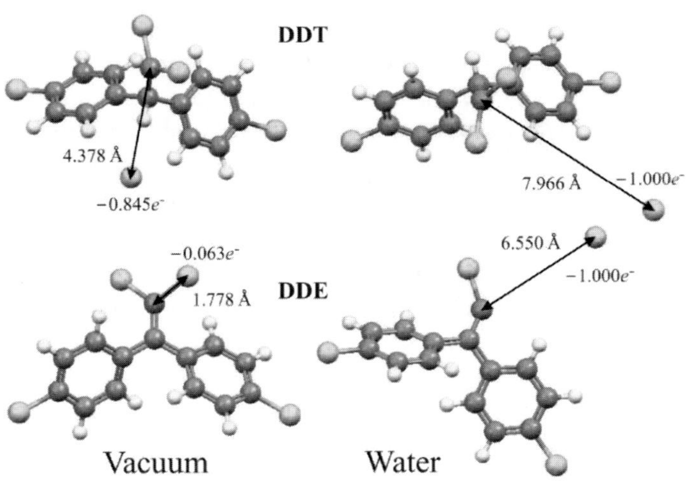

Figure 8. B3LYP/6-31+G(d) optimized geometries of DDT^{-} and DDE^{-} in vacuum and in water.

Even simpler appears the DEA spectrum of DDE (Figure 5), where only the Cl^- negative fragment (m/z = 35) was detected, with a maximum close to zero energy. In agreement, the calculations predict no energy threshold for loss of a chloride anion from the ethene group of DDE^-. A metastable ion peak was detected at an apparent m*/z = 35×35/318 ≈ 3.9, indicating the occurrence of slow (>10 μs) elimination of Cl^- from DDE^- formed at thermal electron energy in the field free region of the spectrometer, as reported in Figure 6.

Relaxed molecular negative ions and electron detachment time. Long-lived (microseconds) molecular negative ions were detected in the DEA spectrum of DDE (m/z = 318), whereas in DDT the corresponding signal at m/z = 354 is absent. Long-lived molecular anions are usually associated with anion states electronically relaxed into their ground state, but vibrationally highly excited [50-53]. The electron detachment time from the molecular anions of DDE was experimentally estimated to be 115 μs at zero energy (see Figure 7). The times of flight of the DDT and DDE molecular anions from the collision cell to the detector are calculated to be, respectively, 43 μs and 41 μs in the DEAS equipment employed, so that the electron detachment time from DDT^- is expected to be < 43 μs. The absence of long-lived molecular anions in the spectrum of DDT could be due to rapid dissociation of a C–Cl bond to give a Cl^- fragment anion and a neutral radical. In agreement, the EA_a calculated for DDT is considerably higher than that of DDE (see Table 2), so that the DDT^- molecular anions formed at zero energy store a large excess of vibrational energy. The absence of metastable peaks (due to slow dissociation processes) in the DEA spectra of DDT is also consistent with a rapid dissociation of its parent molecular anions. In addition, according to the calculated optimized geometries, even in the gas phase the adiabatic molecular anions of DDT are species with an almost dissociated C–Cl bond, where the chlorine atom possesses a large negative charge, as indicated in Figure 8. Indeed, one of the C–Cl bond distances of the geometrically relaxed DDT anion is calculated to be approx. 4.4 Å, with a charge of –0.85*e* on the chlorine atom. In contrast, the calculations predicted a much shorter C-Cl bond length in the gas-phase adiabatic DDE^- molecular anion, where long-lived molecular anions are detected.

Negative ion states in water. PCM calculations were carried out to simulate the effects of a solvent (water) on the energies and geometries of the molecular anions. The stabilization of the solvated adiabatic molecular anions (relative to the corresponding solvated neutral molecules) is sizably larger (> 2 eV) than that calculated in the gas phase (Table 2). This EA_a increase is accompanied by considerable geometrical changes (Figure 8). In particular, according to the calculations, in the solvated anions (including DDE^-) one of the C–Cl bond distances further increases, as well as the negative charge of the chlorine atom, the molecular anions looking more like a neutral radical plus a dissociated chloride anion. As expected, the stabilization of the negative ions on going from the gas phase to water (for instance, 2.992 eV for Cl^-) is calculated to be much larger than that of the neutral state molecules (about 0.2-0.3 eV). As a result, not only the solvent increases EA_v and EA_a, but also stabilizes the fragments produced by DEA by more than 2 eV relative to the gas phase. These results are in qualitative agreement with experimental findings on the stabilization of aromatic negative ions caused by surrounding water molecules [78].

DISCUSSION

It should be preliminary noted that in both the two mechanisms discussed below for e^- attachment to DDT and DDE in cellular ambient, namely, (i) when the source of electrons is suggested to originate from the catalytic cycle of cytochrome P450 enzymes or (ii) when electrons are expected to derive from the mitochondrial electron transport chain, a competition with cellular diatomic oxygen for quasi-free electron attachment is to be considered. On energetic grounds, cellular electrons can be transferred *in vivo* to species with electron affinities larger than that of O_2 (0.45 eV in the gas phase [79]) as discussed earlier [80].

Acquiring an electron in the active center of enzymes. A possible involvement of the DEA process in the metabolic conversion of xenobiotics by cytochrome P450 enzymes has been discussed for a series of non-steroidal anti-inflammatory drugs [43]. Detoxication of such alien species, i.e., the so-called phase I metabolism consisting in cleavage of xenobiotic molecules by cytochrome P450 enzymes, mainly occurs in the hepatocytes under the action of membrane-anchored proteins [81-83]. The P450 cycle catalyzes principally three types of reactions, hydroxylation, epoxidation and oxidation, occurring via a sequential two-electron reduction with involvement of molecular oxygen and protons, although some side reactions are also possible through the P450 cycle [81, 82]. Substrate binding to the P450 active sites shifts the redox potential of heme iron, thus making possible the transfer of the first electron from the redox partner [84]. Transfer of the second electron is believed [84, 85] to reduce the oxy-P450 complex, which is the last relatively stable intermediate of the whole P450 cycle, to form a peroxo-ferric intermediate [82].

The structure of the oxygen complex activated by the second electron, the exact mechanism of oxygen cleavage, and even the mechanism of electron transfer between the components of the P450 system are not well understood yet [86]. In fact, the intermediates formed after involvement of the second electron in the P450 cycle are highly reactive, so that direct experimental evaluation of these stages presents difficulties [84, 85]. Provided that the electron affinity of the substrate molecule exceeds that of molecular oxygen, it could be plausibly suggested that the second electron will interact with a substrate weakly bound in the active center by hydrogen bonds, not with an oxygen molecule. In this case the reactive species could in principle be produced from a substrate "activated" by the second electron via DEA. It is to be noted that a striking regio- and stereoselectivity of the P450 action [84] could be related to the site-selective mechanism of DEA [50-53] where electron attachment into a particular MO leads to a defined set of negatively charged and neutral fragments, depending on the orbital localization properties as well as on energetic conditions.

In the present case we note that both the main metabolites (DDE and DDD) of DDT have counterparts in the gas-phase DEA fragments. This finding is in line with the previously proposed [43] existence of a "DEA-related route" in the P450 catalytic cycle and with the activity of DDT as a P450 inducer [87]. Indeed, the observed elimination of HCl from DDT^{-} leads directly to formation of a negative ion of the first abundant DDT metabolite, i.e., DDE^{-}. The neutral DDD molecule can be formed via a two-stage process. First, dissociation of the DDT^{-} parent anion generates a Cl^{-} negative fragment and the corresponding dechlorinated neutral radical, $[DDT - Cl]^{\bullet}$. Then the latter can abstract a hydrogen atom from, for instance, neighboring lipids, leading to formation of a neutral DDD molecule. In fact, according to B3LYP/6-31+G(d) calculations, the affinity of the $[DDT - Cl]^{\bullet}$ radical for a neutral H atom is 4.25 eV, 3.90 eV when zero-point vibrational energy (ZPE) correction is accounted for, close to the threshold value (3.99 eV) reported by Gregory [30] to disrupt lipid bilayers. It is to be mentioned that the same mechanism of formation of the main metabolite of the model toxicant carbon tetrachloride, namely chloroform, was earlier attributed to dechlorination of CCl_4 by DEA, followed by abstraction of a H atom to form the CCl_3H species in living cells [30]. The affinity (4.89 eV, 4.54 eV with ZPE-correction) of the $[DDE - Cl]^{\bullet}$ radical for a H atom exceeds sizably the threshold value. Tight covalent binding of the reactive dechlorinated radicals in the active centers of cytochrome P450 enzymes is likely to lead to deactivation of the intrinsic system of detoxification of adverse alien species, thus favoring disruption of cells by xenobiotic poisons. Indeed, multiple side reactions are possible through the P450 cycle, and some of them really occur under physiological conditions [86]. Oxidative metabolism of some specific drugs was found to cause a suicide inactivation of the cytochrome P450 system, thus interfering with essential physiological processes [84].

Interaction with electrons "leaked" from mitochondrial respiratory chain. Electron transfer through the mitochondrial respiratory chain generates cellular energy, but is potentially dangerous because reactive oxygen species (ROS) are produced as byproducts starting from one-electron reduction of molecular oxygen, thus generating superoxide anions which can lead to oxidative stress conditions [88]. Mitochondrial failure is considered as the main reason for many human diseases including Parkinson's disease, Alzheimer's disease, cardiac disease, cancer and aging [89, 90], and is usually associated with generation of ROS in the intermembrane space or in the matrix of mitochondria stimulated by "electron leak" from the respiratory chain [91, 92]. Being effective oxidative agents, ROS are believed to cause disruption of mitochondrial membranes through lipid peroxidation and also give rise to mutations of mitochondrial DNA [93, 94]. However, generation of ROS is a natural process, normally occurring in living cells, which also produces several beneficial effects as, for instance, oxidative damage of pathologies, and participates to cellular signaling pathways [88]. Healthy organisms are able to implement self-defense mechanisms via generation of natural antioxidant species or through enzymatic protection that neutralizes ROS and prevents development of the oxidative stress [6, 10]. Thus the existing model of mitochondrial failure based on ROS activity implies the occurrence of a pronounced perturbation of the normal redox balance.

Because the outer mitochondrial membrane is easily accessible to molecules weighting less than 1.5 kDalton [90], it may be concluded that small xenobiotic molecules, like DDT and DDE, can easily penetrate into the intermembrane space. Therefore the proposed mechanism is expected to be particularly important for the organs with high power consumption, i.e., liver, heart, brain, where most mitochondria are located. Indeed, xenobiotic species interacting with the respiratory chain have been associated with ROS formation inside mitochondria [95, 96]. ROS are formed as natural byproducts of the normal oxygen metabolism starting from production of superoxide via oxygen reduction by electrons "leaked" from the respiratory chain [97-99]. It is therefore suggested that DDT and DDE can interact similarly with these electrons inside the intermembrane space. Indeed, provided that the electron affinity of the alien species is close to (or exceeds) that of molecular oxygen, electron transfer to these species becomes possible in the presence of oxygen [80].

The calculated gas-phase EA_as of the compounds under investigation are much larger than that of molecular oxygen (see Table 2). Because all DEA fragments are formed below 1.5 eV in the gas-phase (Figure 5) it is likely that the final states of the products can be accessible via transfer of solvated electrons to DDT and DDE by electron tunneling [100], producing fragments similar to those formed in the gas phase [61, 62]. Indeed, weakly bound (1.0-1.5 eV) prehydrated electrons in water were experimentally found to be involved in dissociative electron transfer reactions producing reductive DNA damage via resonance formation of temporary negative ions [101-103]. In the condensed state this process can be very efficient because of the large increase (one-two orders of magnitude) of the DEA cross-section on going from the gas to the condensed phase [104-106] and when the attaching molecule is embedded in a water cluster [107].

Abstraction of a H atom from lipid bilayers by dechlorinated radicals formed by DEA to DDT and DDE in cellular ambient can initiate peroxidation chain reactions, in agreement with the reported DDT destructive activity directed against mitochondrial membranes [15, 16]. Indeed, the affinity of the $[DDT - Cl]^{\bullet}$ and $[DDE - Cl]^{\bullet}$ radicals for a H atom is calculated to exceed Gregory's threshold, so that also these radicals are able to initiate peroxidation of lipid biomembranes. Disruption of the mitochondrial outer membrane by dechlorinated radicals likely formed in the intermembrane space can result in release of cytochrome c into the intracellular fluid cytosol, thus signaling the whole cell to apoptosis. In fact, apoptosis has been found to proceed through a stage of lipid peroxidation [108, 109]. Attack and disruption of the inner membrane opens the access to the mitochondrial genome located in the matrix. Formation of DNA-adducts, i.e., covalently modified bases, by reactive DEA species can lead to mutations of the mitochondrial genome and a variety of associated disease conditions [110].

Another possible activity of DEA species can be directed against membrane–attached enzymes of respiratory complexes, provided that xenobiotic species can accept an electron from these enzymes. A similar molecular mechanism could lead to inactivation of the respiratory Complex II [17, 18] if the latter can release an electron to DDT or DDE in the active center. Finally, exposure to DDT and DDE results in their penetration into brain tissues [111-114] that can cause modulation of the functionalities of brain mitochondria by withdrawing electrons from their electron-transport chain and generation active species by DEA that, in turn, can be responsible for influencing and alternating brain structure and functions by DDT or, in other words, for observable neurotoxic effects (hyperexcitability, tremor and convulsions) [115-116]. In fact, DDT influence on brain metabolism is well-known [117], significant changes in brain structure are due to DDT exposure [115], and, more specifically, intoxication symptoms in mammals were attributed to disruption in functions of the respiratory chain associated with the uncoupling of oxidative phosphorylation in brain mitochondria [116]. The proposed mechanism of DDT and DDE activity in brain tissues should be responsible for mitochondrial disorder usually associated with oxidative stress [117] but also likely caused by alien active radical species formed by DEA to chlorinated xenobiotics. Such conditions in turn should play a crucial role in the pathogenesis of neurodegenerative diseases [118].

CONCLUSION

Once introduced in living tissues, DDT and DDE can be considered as unnatural electron acceptors possessing high electron-withdrawing ability (electron affinities higher than that of molecular oxygen), being thus able to compete with O_2 for the capture of cellular electrons in active centers of enzymes or near sites of electron "leakage" in the mitochondrial respiratory chain. DEA to DDT and DDE in vacuo efficiently generates very reactive dechlorinated radicals at incident electron energies below 1.5 eV. It can therefore be concluded that the same fragment species are likely formed via dissociative interaction with cellular quasi-free electrons either at interface boundaries (lipid-cytosol-protein) near the mitochondrial electron transport chain or in the hydrophobic active centers of enzymes whose catalytic activity is associated with electron transfer processes.

Possible pathways leading to mitochondrial dysfunctions can be associated with the presence of DDT and DDE in living cells. Their detoxification takes place mainly in hepatocytes, containing a large amount of mitochondria because of the great energy consumption of these cells. This would thus expose a large number of mitochondria to these compounds, depending on the level of intoxication. The occurrence of DEA reactions in the intermembrane space of mitochondria, in the presence of "leaked" electrons and DDT and DDE molecules, can produce foreign bodies in the cell. These neutral and negative fragments are reactive species and can give rise to a cascade of complex biochemical processes in the cell. Some species (including free radicals and DNA adducts) formed by DEA to the halogenated compounds considered here are indeed expected to be dangerous for mitochondrial functionality, leading to membrane peroxidation and DNA mutation. Here we outline possible molecular mechanisms likely leading to mitochondrial failure, that is, a very deleterious effect because mitochondria are recognized as key factors in many areas of biomedical science.

Another aspect of the harmful activity of DDT and DDE in living cells is connected with the ability of their dechlorinated radicals formed by DEA to inactivate normal enzymatic processes by tight covalent binding in the active centers of some enzymes. For instance, in the case of cytochrome P450 system such processes likely lead to suicide deactivation of P450 enzymes, in other words, to inability of the organism to deactivate toxic species, thus becoming vulnerable to harmful xenobiotics. The well-established neurotoxicity caused by DDT can be attributed to its ability to penetrate brain tissues, where a lot of mitochondria will therefore be exposed to these molecules. As a result the mitochondria are expected to lose their functions because of likely disruption of their lipid membranes by dechlorinated radicals that can lead to neuronal apoptosis and, in turn, to observable neurotoxic effects. Of course, the biochemistry of halogenated toxic species is very complex and multifaceted, depending on the difference in absorption, metabolism or excretion of these chemicals by various organisms. However there is no doubt that some initial events of their adverse activity can be attributed to elementary chemical reaction, that is, the interaction with cellular quasi-free electrons via the DEA mechanism, which leads to production of active species as well as withdrawal of electrons from their usual pathways in the living cells.

ACKNOWLEDGMENTS

The authors acknowledge Saint-Petersburg State University (research grant 11.38.219.2014), the Russian Foundation for Basic Research (grants #15-29-05786, #15-02-02809) and the Italian Ministero dell'Istruzione, dell'Università e della Ricerca for financial support.

REFERENCES

[1] Metcalf, R. L. Century of DDT. *J. Agricult. Food Chem.* 1973, *21,* 511-519.

[2] van den Berg, H. Global status of DDT and its alternatives for use in vector control to prevent disease. *Env. Health Persp.* 2009, *117,* 1656-1663.

[3] Corbel V.; N'Guessan R. Distribution, mechanisms, impact and management of insecticide resistance in malaria vectors: a pragmatic review. In *Anopheles mosquitoes: new insights into malaria vectors;* Sylvie M.; Ed.; Intech, 2013, pp. 579-633.

[4] Walker, K. Cost-comparison of DDT and alternative insecticides for malaria control. *Med. Veterin. Entomol.* 2000, *14,* 345-354.

[5] Boul, H. L. DDT residues in the environment – a review with a New Zealand perspective. *New Zealand J. Agricult. Res.* 1995, *38,* 257-277.

[6] Turusov, V.; Rakitsky, V.; Tomatis, L. Dichlorodiphenyltrichloroethane (DDT): ubiquity, persistence, and risks. *Env. Health Persp.* 2002, *110,* 125-128.

[7] Tiemann, U. In vivo and in vitro effects of the organochlorine pesticides DDT, TCPM, methoxychlor, and lindane on the female reproductive tract of mammals: a review. *Reproduct. Toxicol.* 2008, *25,* 316-326.

[8] Jaga, K.; Dharmani, C. Global surveillance of DDT and DDE levels in human tissues. *Int. J. Occupat. Med. Env. Health* 2003, *16,* 7-20.

[9] United Nations Environment Program, Guidance on the Global Monitoring Plan for Persistent Organic Pollutants – Preliminary Version, UNEP: Geneva, Switzerland, 2007.

[10] Gold, B.; Brunk, G. Metabolism of 1, 1, 1-trichloro-2, 2-bis (p-chlorophenyl) ethane and 1, 1-dichloro-2, 2-bis (p-chlorophenyl) ethane in the mouse. *Chem.-Biol. Interact.* 1982, *41,* 327-339.

[11] Kallman, B. J.; Andrews, A. K. Reductive dechlorination of DDT to DDD by yeast. *Science* 1963, *142,* 655-655.

[12] Peterson, J. E.; Robison, W. H. Metabolic products of p, p'-DDT in the rat. *Toxicol. Appl. Pharmacol.* 1964, *6,* 321-327.

[13] Kitamura, S.; Shimizu, Y.; Shiraga, Y.; Yoshida, M.; Sugihara, K.; Ohta, S. Reductive Metabolism of p,p'-DDT and o,p'-DDT by Rat Liver Cytochrome P450. *Drug Metabol. Disposit.* 2002, *30,* 113-118.

[14] Kelner, M.J.; Mclenithan, J.C.; Anders, M.W. Thiol stimulation of the cytochrome P-450-dependent reduction of 1,1,1-(trichloro-2,2-bis(p-chlorophenyl)ethane (DDT) to 1,1-dichloro-2,2-bis(p-chlorophenyl)ethane (DDD). *Biochem. Pharmacol.* 1986, *35*, 1805-1807.

[15] Fiorini, C.; Gilleron, J.; Carette, D.; Valette, A.; Tilloy, A.; Chevalier, S.; Segretain D.; Pointis, G. Accelerated internalization of junctional membrane proteins (connexin 43, N-cadherin and ZO-1) within endocytic vacuoles: an early event of DDT carcinogenicity. *Biochim. Biophys. Acta Biomembran.* 2008, *1778,* 56-67.

[16] Byczkowski, J. Z. Biochemical and ultrastructural changes in biological membranes induced by p,p'-DDT (chlorophenotanum) and its metabolites, *Pol. J. Pharmacol. Pharm.* 1977, *29,* 411-417.

[17] Khan, H. M.; Cutkomp, L. K. Effects of DDT, DDE, and PCBs on mitochondrial respiration. *Bull. Env. Contamin. Toxicol.* 1982, *29,* 577-585.

[18] Nishihara, Y.; Utsumi, K. Effects of 1, 1, 1-trichloro-2, 2-bis-(p-chlorophenyl) ethane (DDT) on ATPase-linked functions in isolated rat-liver mitochondria. *Food Chem. Toxicol.* 1985, *23,* 599-602.

[19] Smith, A. G. Chlorinated hydrocarbon insecticides, in *Handbook of Pesticides Toxicology*; Hayes, W. J.; Laws, E. R. Ed; San Diego/New York: Academic Press Inc., 1991; pp 731-915.

[20] Tsushimoto, G.; Chang, C. C.; Trosko, J. E.; Matsumura, F. Cytotoxic, mutagenic, and cell-cell communication inhibitory properties of DDT, lindane, and chlordane on Chinese hamster cellsin vitro. *Arch. Env. Contamin. Toxicol.* 1983, *12,* 721-729.

[21] Ruch, R. J.; Klaunig, J. E. Effects of tumor promoters, genotoxic carcinogens and hepatocytotoxins on mouse hepatocyte intercellular communication. *Cell Biol. Toxicol.* 1986, *2,* 469-483.

[22] Binelli, A.; Riva, C.; Cogni, D.; Provini, A. Genotoxic effects of p, p′-DDT (1, 1, 1-trichloro-2, 2-bis-(chlorophenyl) ethane) and its metabolites in Zebra mussel (D. polymorpha) by SCGE assay and micronucleus test. *Env. Molec. Mutagenesis* 2008, *49,* 406-415.

[23] Recknagel, R. O.; Glende, E. A.; Dolak, J. A.; Waller, R. L. Mechanisms of carbon tetrachloride toxicity. *Pharmacol. Therap.* 1989, *43,* 139-154.

[24] Weber, L. W.; Boll, M.; Stampfl, A. Hepatotoxicity and mechanism of action of haloalkanes: carbon tetrachloride as a toxicological model. *Crit. Rev. Toxicol.* 2003, *33,* 105-136.

[25] Manibusan, M. K.; Odin, M.; Eastmond, D. A. Postulated carbon tetrachloride mode of action: a review. *J. Environ. Sci. Health C* 2007, *25,* 185-209.

[26] Fujii, K.; Miki, N.; Sugiyama, T.; Morio, M.; Yamano, T.; Miyake, Y. Anaerobic dehalogenation of halothane by reconstituted liver microsomal cytochrome P-450 enzyme system. *Biochem. Biophys. Res. Commun.* 1981, *102,* 507-512.

[27] Kubic, V. L.; Anders, M. W. Mechanism of the microsomal reduction of carbon tetrachloride and halothane. *Chem.-Biol. Interact.* 1981, *34,* 201-207.

[28] Nastainczyk, W.; Ahr, H. J.; Ullrich, V. The reductive metabolism of halogenated alkanes by liver microsomal cytochrome P450. *Biochem. Pharmacol.* 1982, *31,* 391-396.

[29] Brattin, W. J., Glende, E. A.; Recknagel, R. O. Pathological mechanisms in carbon tetrachloride hepatotoxicity. *J. Free Radic. Biol. Med.* 1985, *1,* 27-38.

[30] Gregory, N. L. Carbon tetrachloride toxicity and electron capture. *Nature* 1966, 212, 1460-1461.

[31] Shi, J.; Aisaki, K.; Ikawa, Y.; Wake, K. Evidence of hepatocyte apoptosis in rat liver after the administration of carbon tetrachloride. *Am. J. Pathology* 1998, *153,* 515-525.

[32] Slater, T. F.; Cheeseman, K. H.; Ingold, K. U.; Rice-Evans, C.; Sies, H. Carbon Tetrachloride Toxicity as a Model for Studying Free-Radical Mediated Liver Injury [and Discussion]. *Phil. Trans. Royal Soc. London B: Biol. Sci.* 1985, *311,* 633-645.

[33] Bauer, A.; Schumann, A.; Gilbert, M.; Wilhelm, C.; Hengstler, J. G.; Schiller, J.; Fuchs, B. Evaluation of carbon tetrachloride-induced stress on rat hepatocytes by 31 P NMR and MALDI-TOF mass spectrometry: lysophosphatidylcholine generation from unsaturated phosphatidylcholines. *Chem. Phys. Lipids* 2009, *159,* 21-29.

[34] Boll, M.; Weber, L. W.; Becker, E.; Stampfl, A. Pathogenesis of carbon tetrachloride-induced hepatocyte injury bioactivation of CCl4 by cytochrome P450 and effects on lipid homeostasis. *Zeitschrift für Naturforschung* C 2001, *56,* 111-121.

[35] Poli, G.; Albano, E.; Dianzani, M. U. The role of lipid peroxidation in liver damage. *Chem. Phys. Lipids* 1987, *45,* 117-142.

[36] Lovelock, J. E. Affinity of organic compounds for free electrons with thermal energy: its possible significance in biology, *Nature* 1961, *189,* 729-732.

[37] Lovelock, J. E.; Zlatkis, A.; Becker, R. S. Affinity of polycyclic aromatic hydrocarbons for electrons with thermal energies: its possible significance in carcinogenesis. *Nature* 1962, *193,* 540-541.

[38] Lovelock, J. E.; Simmonds, P. G.; VandenHeuvel, W. A. Affinity of steroids for electrons with thermal energies. *Nature* 1963, *197,* 249-251.

[39] Antonello, S.; Maran, F. Intramolecular dissociative electron transfer. *Chem. Soc. Rev.* 2005, *34,* 418-428.

[40] Maran, F.; Workentin, M. S. Dissociative electron transfer. *Interf.-Electrochem. Soc.* 2002, *11,* 44-50.

[41] Marcus, R. A. Electron transfer reactions in chemistry. Theory and experiment. *Rev. Mod. Phys.* 1993, *65,* 599-610.

[42] Pshenichnyuk, S. A.; Lomakin, G. S.; Modelli, A. Degradation of gas phase decabromodiphenyl ether by resonant interaction with low-energy electrons. *Phys. Chem. Chem. Phys.* 2011, *13,* 9293-9300.

[43] Pshenichnyuk, S. A.; Modelli, A. Electron attachment to antipyretics: Possible implications of their metabolic pathways. *J. Chem. Phys.* 2012, *136,* 234307.

[44] Pshenichnyuk, S. A.; Modelli, A. Can mitochondrial dysfunction be initiated by dissociative electron attachment to xenobiotics? *Phys. Chem. Chem. Phys.* 2013, *15,* 9125-9135.

[45] Modelli, A.; Pshenichnyuk S. A., Gas-phase dissociative electron attachment to flavonoids and possible similarities to their metabolic pathways. *Phys. Chem. Chem. Phys.* 2013, *15,* 1588-1600.

[46] Asfandiarov, N. L.; Pshenichnyuk, S. A.; Vorob'ev, A. S.; Nafikova, E. P.; Elkin, Y. N.; Pelageev, D. N.; Koltsova, E. A.; Modelli, A. Electron attachment to some naphthoquinone derivatives: long-lived molecular anion formation. *Rapid Commun. Mass Spectrom.* 2014, *28,* 1580-1590.

[47] Pshenichnyuk, S. A.; Modelli, A. Resonance electron attachment to plant hormones and its likely connection with biochemical processes. *J. Chem. Phys.* 2014, *140,* 034313.

[48] Pshenichnyuk, S. A.; Komolov, A. S. Dissociative electron attachment to anthralin to model its biochemical reactions. *J. Phys. Chem. Lett.* 2014, *5,* 2916-2921.

[49] Pshenichnyuk, S. A.; Komolov, A. S. Dissociative electron attachment to resveratrol as a likely pathway for generation of the H_2 antioxidant species inside mitochondria. *J. Phys. Chem. Lett.* 2015, *6,* 1104-1110.

[50] Schulz, G. J. Resonances in electron impact on diatomic molecules. *Rev. Mod. Phys.* 1973, *45,* 423-486.

[51] Illenberger, E.; Momigny, J. *Gaseous molecular ions. An introduction to elementary processes induced by ionization*, Steinkopff Verlag Darmstadt, Springer-Verlag, New York, 1992.

[52] Christophorou, L. G. *Electron-molecule interactions and their applications*, Academic Press, Orlando, 1984.

[53] Khvostenko, V. I. *Negative ions mass spectrometry in organic chemistry*, Moscow, Nauka, 1981 (in Russian).

[54] Sanche, L.; Schulz, G. J. Electron transmission spectroscopy: Rare gases. *Phys. Rev. A* 1972, *5(4),* 1672-1683.

[55] Allan, M. Study of triplet states and short-lived negative ions by means of electron impact spectroscopy. *J. Elec. Spec. Rel. Phenom.* 1989, *48*, 219-351.

[56] Komolov, A. S.; Lazneva, E. F.; Akhremtchik S.N.; Chepilko N. S.; Gavrikov A.A. Unoccupied Electronic States at the Interface of Oligo(phenylene-vinylene) Films with Oxidized Silicon. *J. Phys. Chem. C* 2013, *117*, 12633–12638.

[57] Komolov, A. S.; Lazneva, E. F.; Gerasimova, N. B.; Panina Yu. A.; Baramygin A.V.; Zashikhin G. D.; Pshenichnyuk, S. A. Structure of Vacant Electronic States of an Oxidized Germanium Surface upon Deposition of Perylene Tetracarboxylic Dianhydride Films. *Physics of the solid state* 2016, *58*, 377-381.

[58] Pshenichnyuk, S. A.; Komolov, A. S. Relation Between Electron Scattering Resonances of Isolated NTCDA Molecules and Maxima in the Density of Unoccupied States of Condensed NTCDA Layers. *J. Phys. Chem. A* 2012, *116*, 761-766.

[59] Pshenichnyuk, S. A.; Kukhto, A. V.; Kukhto, I. N.; Komolov, A. S. Spectroscopic States of PTCDA Negative Ions and Their Relation to the Maxima of Unoccupied State Density in the Conduction Band. *Tech. Phys.* 2011, *56*, 754-759.

[60] Christophorou, L. G.; Hadjiantoniou, D. Electron attachment and molecular toxicity. *Chem. Phys. Lett.* 2006, *419,* 405-410.

[61] Abel, B.; Buck, U.; Sobolewski, A. L.; Domcke, W. On the nature and signatures of the solvated electron in water. *Phys. Chem. Chem. Phys.* 2012, *14*, 22-34.

[62] Siefermann, K. R.; Liu, Y.; Lugovoy, E.; Link, O.; Faubel, M.; Buck, U.; Winter, B.; Abel, B. Binding energies, lifetimes and implications of bulk and interface solvated electrons in water. *Nat. Chem.* 2010, *2,* 274-279.

[63] Poully, J. C.; Nieuwjaer, N.; Schermann, J. P. Structure and dynamics of molecules of pharmaceutical interest in gas phase and in aqueous phase. *Physica Scripta* 2008, *78(5),* 058123.

[64] Modelli, A.; Jones, D.; Distefano, G. ETS study of the negative ion states of t-butyl and trimethylsilyl derivatives of ethylene and benzene. *Chem. Phys. Lett.* 1982, *86,* 434-437.

[65] Pshenichnyuk, S. A.; Modelli, A. ETS and DEAS studies of the reduction of xenobiotics in mitochondrial intermembrane space, In *Mitochondrial Medicine: Volume II, Manipulating Mitochondrial Function, Methods in Molecular Biology*, Weissig, V. ; Edeas M. , Ed.; 2015, vol. 1265, Springer Science + Business Media New York, pp. 285-305.

[66] Edelson, D.; Griffiths, J. E.; McAfee Jr, K. B. Autodetachment of electrons in sulfur hexafluoride. *J. Chem. Phys.* 1962, *37,* 917-918.

[67] Pshenichnyuk, S. A.; Vorob'ev, A. S.; Asfandiarov, N. L.; Modelli, A. Molecular anion formation in 9, 10-anthraquinone: Dependence of the electron detachment rate on temperature and incident electron energy. *J. Chem. Phys.* 2010, *132,* 244313.

[68] Pshenichnyuk, S. A.; Vorob'ev, A. S.; Modelli, A. Resonance electron attachment and long-lived negative ions of phthalimide and pyromellitic diimide. *J. Chem. Phys.* 2011, *135,* 184301.

[69] Frisch M. J. et al, Gaussian 09, Revision A.02, Gaussian, Inc., Wallingford CT, 2009.

[70] Becke, A. D. Density-functional thermochemistry. III. The role of exact exchange. *J. Chem. Phys.* 1993, *98,* 5648-5652.

[71] Simons, J.; Jordan, K. D. Ab initio electronic structure of anions. *Chem. Rev.* 1987, *87(3),* 535-555.

[72] Staley, S. W.; Strnad, J. T. Calculation of the energies of. pi.* negative ion resonance states by the use of Koopmans' theorem. *J. Phys. Chem.* 1994, *98,* 116-121.

[73] Chen, D.; Gallup, G. A. The relationship of the virtual orbitals of self-consistent-field theory to temporary negative ions in electron scattering from molecules. *J. Chem. Phys.* 1990, *93,* 8893-8901.

[74] Modelli, A. Electron attachment and intramolecular electron transfer in unsaturated chloroderivatives. *Phys. Chem. Chem. Phys.* 2003, *5,* 2923-2930.

[75] Scheer, A. M.; Burrow P. D. π^* orbital system of alternating phenyl and ethynyl groups: Measurements and calculations, *J. Phys. Chem. B* 2006, *110,* 17751-17756.

[76] Komolov, A. S.; Lazneva, E. F.; Pshenichnyuk, S. A.; Chepilko, N. S.; Tomilov, A. A.; Gerasimova, N. B.; Lezov, A. A.; Repin, P. S. Electronic Properties of the Interface Between Hexadecafluoro Copper Phthalocyanine and Unsubstituted Copper Phthalocyanine Films. *Semiconductors* 2013, *47*, 956-961.

[77] Burrow, P. D.; Gallup, G. A.; Modelli, A. Are There π* Shape Resonances in Electron Scattering from Phosphate Groups? *J. Phys. Chem. A* 2008, *112,* 4106-4113.

[78] Lyapustina, S. A.; Xu, S.; Nilles, J. M.; Bowen, K. H. Solvent-induced stabilization of the naphthalene anion by water molecules: A negative cluster ion photoelectron spectroscopic study, *J. Chem. Phys.* 2000, *112,* 6643-6648.

[79] Ervin, K. M.; Anusiewicz, I.; Skurski, P.; Simons, J.; Lineberger, W. C. The only stable state of O_2^- is the X 2Πg ground state and it (still!) has an adiabatic electron detachment energy of 0.45 eV. *J. Phys. Chem. A* 2003, *107,* 8521-8529.

[80] Biaglow, J. E. Cellular electron transfer and radical mechanisms for drug metabolism. *Radiat. Res.* 1981, *86,* 212-242.

[81] Denisov, I. G.; Makris, T. M.; Sligar, S. G.; Schlichting, I. Structure and chemistry of cytochrome P450. *Chem. Rev.* 2005, *105,* 2253-2278.

[82] Rittle, J.; Green, M. T. Cytochrome P450 compound I: capture, characterization, and CH bond activation kinetics. *Science* 2010, *330,* 933-937.

[83] Miners, J. O.; Birkett, D. J. Cytochrome P4502C9: an enzyme of major importance in human drug metabolism. *Br. J. Clinic. Pharm.* 1998, *45,* 525-538.

[84] Ortiz de Montellano, P. R. Hydrocarbon hydroxylation by cytochrome P450 enzymes. *Chem. Rev.* 2010, *110,* 932-948.

[85] Meunier, B.; De Visser, S. P.; Shaik, S. Mechanism of oxidation reactions catalyzed by cytochrome P450 enzymes. *Chem. Rev.* 2004, *104,* 3947-3980.

[86] Bernhardt, R. Cytochrome P450: structure, function, and generation of reactive oxygen species. In *Reviews of Physiology Biochemistry and Pharmacology*, 1995, Vol.127, Springer Berlin Heidelberg, pp. 137-221.

[87] Nims, R. W.; Lubet, R. A.; Fox, S. D.; Jones, C. R.; Thomas, P. ; Reddy, A. B.; Kocarek, T. A. Comparative pharmacodynamics of CYP2B induction by DDT, DDE, and DDD in male rat liver and cultured rat hepatocytes, *J. Toxicol. Environ. Health.* 1998, *53,* 455-477.

[88] Murphy, M.P. How mitochondria produce reactive oxygen species, *Biochem. J.* 2009, *417,* 1-13.

[89] Koopman, W. J. H.; Willems, P. H. G. M.; Smeitink, J. A. M. Monogenic mitochondrial disorders, *N. Engl. J. Med.* 2012, *366,* 1132-1141.

[90] Szewczyk, A.; Wojtczak, L. Mitochondria as a pharmacological target, *Pharmacol. Rev.* 2002, *54,* 101-127.

[91] Andreyev, A. Y.; Kushnareva, Y. E.; Starkov, A. A. Mitochondrial metabolism of reactive oxygen species. *Biochemistry (Moscow)* 2005, *70,* 200-214.

[92] Selivanov, V. A.; Votyakova, T. V.; Pivtoraiko, V. N.; Zeak, J.; Sukhomlin, T.; Trucco, M.; Roca, J.; Cascante, M. Reactive oxygen species production by forward and reverse electron fluxes in the mitochondrial respiratory chain. *PLoS Comput Biol* 2011, *7,* e1001115.

[93] Costa, R. A. P.; Romagna, C. D.; Pereira, J. L.; Souza-Pinto, N. C. The role of mitochondrial DNA damage in the citotoxicity of reactive oxygen species. *J. Bioenerg. Biomembran.* 2011, *43,* 25-29.

[94] Greaves, L. C.; Reeve, A. K.; Taylor, R. W.; Turnbull, D. M. Mitochondrial DNA and disease. *J. Pathology* 2012, *226,* 274-286.

[95] Turrens, J. F. Mitochondrial formation of reactive oxygen species , *J. Physiol.* 2003, *552,* 335-344.

[96] Scatena, R.; Bottoni, P.; Botta, G.; Martorana, G. E.; Giardina, B. The role of mitochondria in pharmacotoxicology: a reevaluation of an old, newly emerging topic. *Am. J. Physiol. Cell Physiol.* 2007, *293,* C12-C21.

[97] Liu, Y. ; Fiskum, G.; Schubert, D. Generation of reactive oxygen species by the mitochondrial electron transport chain, *J. Neurochem.* 2002, *80,* 780-787.

[98] Lenaz, G. The mitochondrial production of reactive oxygen species: mechanisms and implications in human pathology, *IUBMB Life* 2001, *52,* 159-164.

[99] Chen, Q.; Vazquez, E. J.; Moghaddas, S.; Hoppel, C. L.; Lesnefsky, E. J. Production of reactive oxygen species by mitochondria central role of complex III. *J. Biol. Chem.* 2003, *278,* 36027-36031.

[100] de la Lande, A.; Babcock, N. S.; Řezáč, J.; Lévy, B.; Sanders, B. C.; Salahub, D. R. Quantum effects in biological electron transfer. *Phys. Chem. Chem. Phys.* 2012, *14*, 5902-5918.

[101] Wang, C. R.; Nguyen, J.; Lu, Q. B. Bond breaks of nucleotides by dissociative electron transfer of nonequilibrium prehydrated electrons: a new molecular mechanism for reductive DNA damage. *J. Am. Chem. Soc.* 2009, *131,* 11320-11322.
[102] Ptasińska, S.; Denifl, S.; Scheier, P.; Illenberger, E.; Märk, T. D. Bond-and site-selective loss of H atoms from nucleobases by very-low-energy electrons (<3 eV). *Angew. Chem. Int. Ed.* 2005, *44,* 6941-6943.
[103] Ptasińska, S.; Sanche, L. Dissociative electron attachment to hydrated single DNA strands. *Phys. Rev. E* 2007, *75,* 031915/1-5.
[104] Alizadeh, E.; Sanche, L. Precursors of Solvated Electrons in Radiobiological Physics and Chemistry, *Chem. Rev.* 2012, *112,* 5578-5602.
[105] Sambe, H.; Ramaker, D. E.; Deschenes, M.; Bass, A. D.; Sanche, L. Absolute cross section for dissociative electron attachment in O 2 condensed on Kr film. *Phys. Rev. Lett.* 1990, *64,* 523-526.
[106] Lu, Q.-B.; Sanche, L. Large enhancement in dissociative electron attachment to HCl adsorbed on H_2O ice via transfer of presolvated electrons, *J. Chem. Phys.* 2001, *115,* 5711-5713.
[107] Fabrikant, I. I.; Caprasecca, S.; Gallup, G. A.; Gorfinkiel, J. D. Electron attachment to molecules in a cluster environment, *J. Chem. Phys.* 2012, *136,* 184301-8.
[108] Ow, Y.-L. P.; Green, D. R.; Hao, Z.; Mak, T. W. Cytochrome c: functions beyond respiration, *Nat. Rev. Mol. Cell Biol.* 2008, *9,* 532-542.
[109] Wang, C.; Youle, R. J. The role of mitochondria in apoptosis, *Annu. Rev. Genet.* 2009, *43,* 95-118.
[110] Wallace, D. C. Diseases of the mitochondrial DNA. *Annu. Rev. Biochem.* 1992, *61,* 1175-1212.
[111] Gingell, R.; Wallcave, L. Species differences in the acute toxicity and tissue distribution of DDT in mice and hamsters. *Toxicol. Appl. Pharmacol.* 1974, *28,* 385-394.
[112] Morrison, G. Penetration of the blood-brain-cerebral spinal fluid barrier by DDT. *Bull. Environ. Contamin. Toxicol.* 1971, *6,* 48-54.
[113] Byczkowski, J. Z. The mode of action of p, p′-DDT on mammalian mitochondria. *Toxicol.* 1976, *6,* 309-314.
[114] Woolley, D. E.; Runnells, A. L. Distribution of DDT in brain and spinal cord of the rat. *Toxicol. Appl. Pharmacol.* 1967, *11,* 389-395.

[115] Iwaniuk, A. N.; Koperski, D. T.; Cheng, K. M.; Elliott, J. E.; Smith, L. K.; Wilson, L. K.; Wylie, D. R. The effects of environmental exposure to DDT on the brain of a songbird: changes in structures associated with mating and song. *Behav. Brain Res.* 2006, *173,* 1-10.
[116] Hrdina, P. D.; Singhal, R. L.; Peters, D. A.; Ling, G. M. Role of brain acetylcholine and dopamine in acute neurotoxic effects of DDT. *Europ. J. Pharmacol.* 1971, *15,* 379-382.
[117] Adam-Vizi, V. Production of reactive oxygen species in brain mitochondria: contribution by electron transport chain and non-electron transport chain sources. *Antiox. Redox Signal.* 2005, *7,* 1140-1149.
[118] Lin, M. T.; Beal, M. F. Mitochondrial dysfunction and oxidative stress in neurodegenerative diseases. *Nature* 2006, *443,* 787-795.

In: DDT
Editor: Kathleen Sanders
ISBN: 978-1-53610-009-9

Chapter 2

DDT AND PYRETHROID RESISTANCE IN *AEDES AEGYPTI* (L.) AND *AEDES ALBOPICTUS* (SKUSE): PAST, PRESENT, AND FUTURE

Hitoshi Kawada*
Institute of Tropical Medicine, Nagasaki University, Nagasaki, Japan

ABSTRACT

Aedes aegypti, an important vector of viral diseases such as yellow fever, dengue, chikungunya, and zika, is thought to have originated in the African continent. The subspecies that has been domesticated and adapted to anthropogenic environments (*Aedes aegypti aegypti*) expanded its habitat around human domiciles and has been dispersed by human movement. Worldwide eradication programs with organochlorine insecticides in 1940s and 1950s may have caused high selection pressure on mosquitoes and expanded the distribution of insecticide-resistant *Ae. aegypti* populations. DDT resistance in *Ae. aegypti* was first reported in Caribbean countries in the 1950s, and that resistance persists in almost all regions that had achieved *Ae. aegypti* eradication, despite the fact that DDT is no longer used.

Aedes albopictus inhabits tropical regions but is also found in temperate areas, including Japan, Europe, America, and Australia. Although dengue hemorrhagic fever has not been prevalent in Japan for

* Corresponding author: Phone: 81-95-819-7811; Fax: 81-95-819-7812; Email address: vergiss@nagasaki-u.ac.jp.

the past 50 years and a chikungunya outbreak has never been reported in Japan, the possibility of viral disease outbreaks exists because of the widespread distribution of the vector *Ae. albopictus*. DDT resistance in this species seems to be common in Japan, while resistance to other insecticides such as pyrethroids, organophosphates, and carbamates is uncommon.

It is noteworthy that there has been a worldwide focus for the past three decades on the dispersal of shipping containers of used tires that carry breeding mosquitoes. The mosquito species that has played the leading part in these events is *Ae. albopictus*, which together with four other Japanese mosquito species is thought to have arrived at the western coastal ports of the United States in used tires by 1983, becoming widely distributed in the United States and Brazil by 1986. Since 1986, tire shipments infested with *Ae. albopictus* have been found in South and Central America, South and West Africa, Oceania, and Europe. Used tire trading occurs worldwide, with complicated commercial networks in many countries including South Africa, Kenya, Uganda, Niger, Nigeria, and Ghana. The history of the worldwide invasion of *Ae. aegypti* in association with human activity may be longer than that of *Ae. albopictus*. The worldwide transportation of vector mosquitoes in association with human activity may result in the introgression of DDT and other insecticide-resistance genes.

DDT has played an important role as the most effective insecticide for indoor residual spraying (IRS). The Stockholm Convention seeks the elimination of DDT. There is a concern, however, that a ban on DDT use could adversely affect the malaria burden, since many countries have relied on the effectiveness and low cost of DDT treatment. The use of DDT is now recommended only where the intervention is appropriate and effective in the local epidemiologic situation. The continued efficacy of DDT, on the other hand, seems to have given rise to DDT resistance and cross-resistance to pyrethroids in *Ae. aegypti* and *Ae. albopictus*, as well as in *Anopheles* mosquitoes.

Keywords: *Aedes aegypti*, *Aedes albopictus*, DDT, mosquito, pyrethroid, resistance

1. Introduction

Insect pests are species that have emerged due to human activity and are harmful or disadvantageous from the human point of view. Sanitary pests and vectors of infectious diseases emerged as humans adopted communal living. Agricultural pests appeared as humans started to cultivate fields and grow crops. In order to ensure and improve human comfort, insecticides have been developed. Environmental persistence with maximum efficacy and minimum toxicity are desirable traits of good insecticides. Minimal impacts on non-target organisms, including fish, insects, algae, and other species, are also desirable. Currently, the probability of inventing a new insecticide is 1 in 100,000; that is, 100,000 useless chemicals are synthesized before one beneficial insecticide is invented. The probability of invention was 100 times greater 60 years ago during the brilliant age of organochloride and organophosphate insecticides. The development cost for one new insecticide today is more than 100 million dollars. Development costs have increased exponentially over the past 60 years, mainly because of the increase in toxicological data requirements. Nearly ten years are required for the development of a single new insecticide. Thousands of chemicals are synthesized and biologically screened every week. It is evident that developing an insecticide is extremely expensive and laborious. It is, therefore, our duty to use insecticides in the most effective and prudent manner possible in order to maintain their effectiveness and sustain their use.

The most important event in the history of mosquito control was the invention of dichloro-diphenyl-trichloroethane (DDT) by Dr. Paul Müller, who received the Nobel Prize for Chemistry in 1948. The long persistence and excellent killing efficacy of DDT as the agent in indoor residual spraying (IRS) are responsible for its brilliant success after the Second World War. Nevertheless, the first DDT resistance case in *Anopheles* mosquitoes was detected several years later. Furthermore, DDT resistance was responsible for the increase in the incidence of malaria in the 1960s (Bruce-Chwatt 1985). Due to the failure of the malaria eradication program with DDT, the World Health Organization (WHO) has altered its policy from eradication to control by primary health care and the use of pyrethroid-impregnated mosquito nets. Nowadays, pyrethroids are emerging as the predominant insecticides for vector control. Globally, pyrethroids constitute 100% of WHO-recommended insecticides for the treatment of long-lasting insecticidal nets (LLINs) and approximately 81% of spray usage, with 68% of the latter constituting residual spraying and 24% space spraying (van der Berg et al. 2012). The use of pyrethroids in insecticide-treated bed nets (ITNs) is a simple and inexpensive protection measure against malaria and has been shown to reduce morbidity in children less than five years old by 50% and global child mortality by 20–30% (Binka et al. 1996, Lengeler et al. 1996, Nevill et al. 1996). Development of LLINs was the most significant breakthrough measure for malaria control (Guillet et al. 2001).

The discovery of the phenoxybenzyl alcohol moiety accelerated the development of photo-stable pyrethroids for outdoor use in agriculture. These "second generation" pyrethroids have been used worldwide as good vector control agents with various application techniques. However, photo-stable and highly effective pyrethroids may accelerate the development of pyrethroid resistance in mosquito populations. Among pyrethroids that are used for vector control, 98.7% contain photo-stable pyrethroids such as α-cypermethrin, bifenthrin, cyfluthrin, cypermethrin, deltamethrin, etofenprox, λ-cyhalothrin, and permethrin (Zaim and Jambulingam 2007). Pyrethroid resistance will be a major problem for vector control programs, since there are currently no suitable chemical substitutes for pyrethroids.

Toxicity, environmental persistence, and bioaccumulation were the principal disadvantages of DDT that led to the global ban (except for vector control use in tropical areas) that was formalized under the Stockholm Convention on Persistent Organic Pollutants. Global use of DDT, however, has not changed substantially since the Stockholm Convention came into force. India has dominated the global use of DDT (>80%), although it showed a modest decline in use after 2005, while in the African region, DDT use increased sharply until 2008 along with the expansion of IRS programs. In 2009, use of DDT in the African region decreased, primarily due to changes in three countries, Ethiopia, Mozambique, and Uganda, because of an increase in DDT resistance by malaria vectors and/or political changes (van den Berg et al. 2012).

It seems that DDT resistance is still concealing a potency of developing pyrethroid resistance in the other vector mosquitoes, such as *Ae. aegypti* and *Ae. albopictus*, as well as in *Anopheles* mosquitoes, since its greater or equivalent fitness costs as compared to those for susceptible strains (Kliot and Ghanim 2012), common metabolic pathway (glutathione transferase), and common target site (voltage-gated sodium channel, VGSC). The resistance problem with DDT, therefore, may have implications for the use of pyrethroids, the most promising alternatives for DDT. In this historical review of DDT resistance in some vector mosquitoes, the present status of resistance in the area the author has investigated, along with anticipated future problems caused by DDT resistance, are discussed.

2. Nationwide Investigation of Pyrethroid Resistance in Mosquitoes in Vietnam

The presence of *Ae. aegypti* was first recorded in Vietnam in 1915 (Stanton 1920), and since then it has maintained hyperendemicity in most cities. Although *Ae. aegypti* is one of the most important disease vector mosquitoes, and many studies have extensively investigated its insecticide resistance, especially in Asian and South and Central American countries, few studies on insecticide resistance in *Ae. aegypti* have been published in Vietnam (Huber et al. 2003, Vu et al. 2004). These studies have indicated that pyrethroid resistance in these vector mosquitoes seems to be high. In the course of the National Dengue Control Program in Vietnam, the ecological differences between two dengue vectors with different geographical origins, namely, *Aedes aegypti* and *Aedes albopictus*, have become a subject of discussion. The ecological differences between the vectors may be the reason for the differences in the epidemics. In order to examine the nationwide distribution of the two species and the distribution of pyrethroid resistance, used tires, one of the main breeding grounds of the two species, were targeted for larval collection. Used tires were suitable targets for our study as standard breeding sites because they are extensively and commonly distributed along roads in Vietnam. Distribution analysis of pyrethroid susceptibility in three major mosquito species, *Ae. aegypti*, *Ae. albopictus*, and *Culex quinquefasciatus* (Say), was conducted in 2006, 2007, and 2008. The relationships between pyrethroid susceptibility, geographical factors, and province-based annual pyrethroid use were also analyzed.

2.1. Geographical Distribution of Mosquito Species in Vietnam

Collection sites were categorized into five areas: (1) northern mountainous area (172–506 m in elevation), (2) northern plains area, (3) eastern coastal area, (4) southern mountainous area (103–1563 m in elevation), and (5) southern wetlands. Among the 19,188 tires located during the investigation, 4,757 were examined and 2,468 of these (51.9%) contained water; 852 (34.5%) contained mosquitoes, including 653 (26.5%) that contained dengue vector mosquitoes (*Ae. aegypti* and *Ae. albopictus*). In total, 8,771 *Ae. aegypti*, 5,916 *Ae. albopictus*, and 11,356 *Cx. quinquefasciatus* larvae were collected. In the northern part of Vietnam, *Ae. albopictus* was dominant, but this dominance gradually declined toward the south. *Ae. aegypti* was dominant in the southern region. In the eastern coastal areas of the southern region, almost 100% of the *Aedes* larvae collected were *Ae. aegypti*. In the mountainous areas, however, the number of *Ae. albopictus* increased, suggesting that the distribution of the two species is determined by different environmental gradients (Figure 1, Kawada et al. 2009a, Higa et al. 2010).

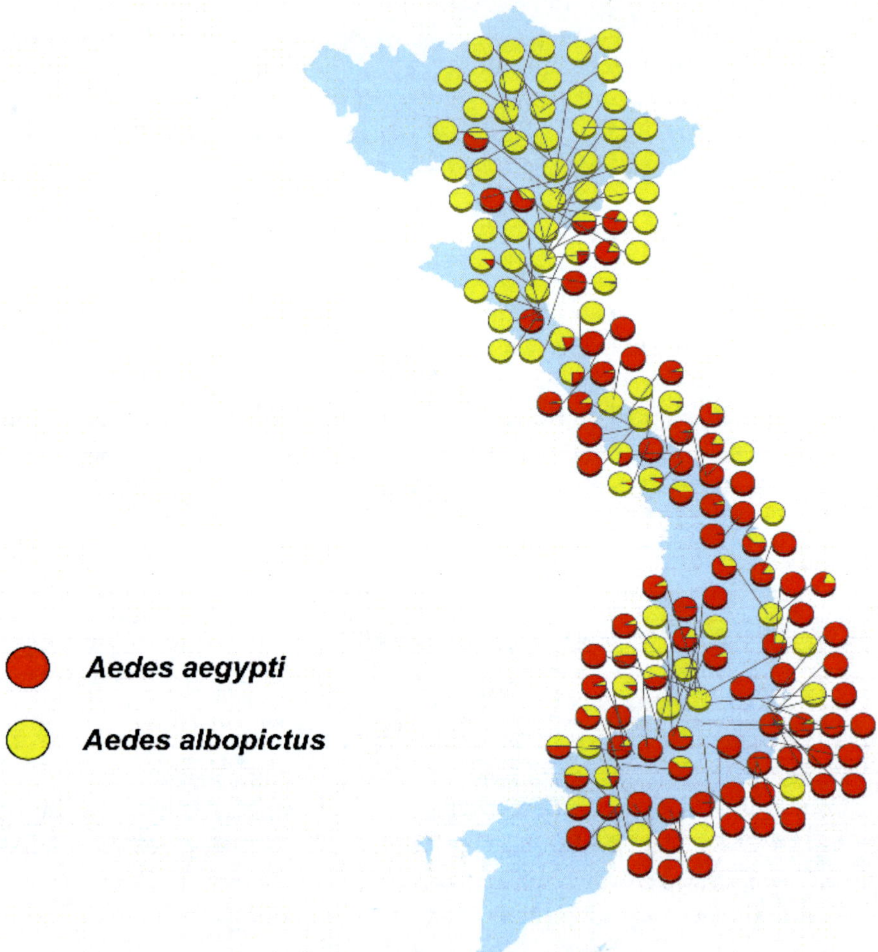

Figure 1. Species composition of *Aedes aegypti* and *Ae. albopictus* larvae collected from used tires in Vietnam (Kawada et al. 2009a, Higa et al. 2010).

2.2. Distribution Analysis of Insecticide Susceptibility in Mosquitoes

The average susceptibility index (an index of 1 indicates the most susceptible and an index of 36 the least susceptible to *d*-allethrin) and the proportion of mosquitoes with susceptibility indices greater than 20 were 24.9 and 59.7% for *Ae. aegypti*, 8.5 and 8.0% for *Ae. albopictus*, and 12.6 and 21.9% for *Cx. quinquefasciatus*, respectively (Figure 2). The reduction in susceptibility was most prominent in *Ae. aegypti*, while the reduction in the susceptibility of *Ae. albopictus* and *Cx. quinquefasciatus* was moderate. The susceptibility of the abovementioned species to *d*-allethrin was lower in southern areas (<13°N) than in northern areas (>13°N) of Vietnam. For *Ae. aegypti*, univariate analysis showed significant increases in the susceptibility indices with decreases in the latitude of the collection points. In contrast, there was no such significant correlation in *Ae. albopictus* and *Cx. quinquefasciatus*. In *Ae. aegypti*, a significant correlation between the susceptibility indices and the total annual pyrethroid use for malaria control during 1998 and 2002 was also observed, while no such significant correlations were seen in *Ae. albopictus* and *Cx. quinquefasciatus* (Kawada et al. 2009a).

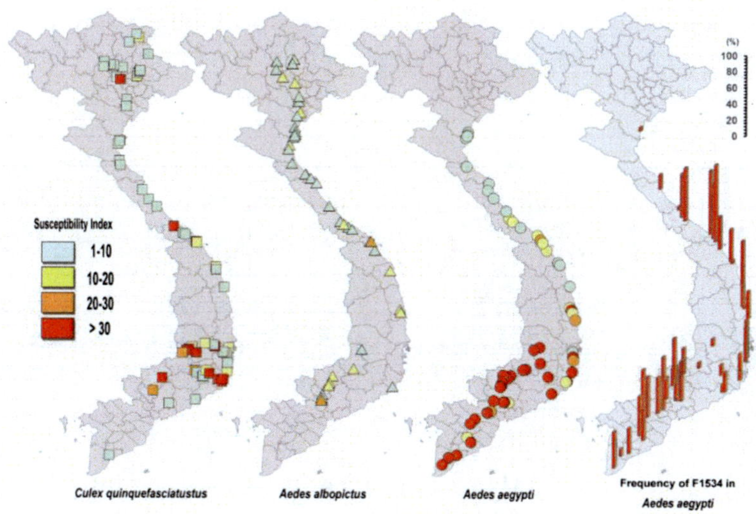

Figure 2. Pyrethroid (*d*-allethrin) susceptibility distribution of mosquito larvae, *Culex quinquefasciatus*, *Aedes albopictus*, and, *Ae. aegypti* collected from used tires in Vietnam, and the distribution of the F1534C point mutation in *Ae. aegypti* (Kawada et al. 2009a, 2009b).

Vu et al. (2004) reported a similar pattern in pyrethroid susceptibility in *Ae. aegypti* in Vietnam. They conducted a WHO standard bioassay and found that these mosquitoes were susceptible to pyrethroids in many places in the North and Central regions but were resistant in the South and Central Highlands in Vietnam. They concluded that this discrepancy in pyrethroid susceptibility in different regions was due to the longer and more extended use of pyrethroids in malaria and dengue fever control as well as in agriculture in the Southern and Central Highlands. In fact, pyrethroids have been extensively used as residual treatments in houses and in pyrethroid-impregnated bed nets for malaria control as a part of the National Malaria Control Program (Verlé et al. 1999, Hung et al. 2002, Nam et al. 2005). Pyrethroid use for malaria control seems to be an important factor in developing pyrethroid resistance in *Ae. aegypti* in the highland region, since the dengue fever and dengue hemorrhagic fever (DF/DHF) cases are not serious in that area (Nguyen 1994), and forest malaria continues to be endemic in the highlands as compared to the other regions (Erhart et al. 2004, 2005, Thang et al. 2008) and consequently the amount of pyrethroid treatment for dengue vector control in the highland region is lower than in other regions. Pyrethroid and DDT treatment for malaria vector control appears to have been intensively conducted in the interior and along the periphery of human habitation areas, where, incidentally, the breeding and resting sites of *Ae. aegypti* are located. This might account for the strong selection pressure on *Ae. aegypti* and the weaker pressure on *Ae. albopictus*, since Asian *Ae. aegypti* is generally a domestic and endophagic species that has a greater preference for indoor breeding than *Ae. albopictus* (Hawley 1988, Edman et al. 1997, Ishak et al. 1997, Higa et al. 2001, Tsuda et al. 2006). Several studies have reported pyrethroid resistance in both Asian *Ae. aegypti* and *Ae. albopictus*. Most of these studies have shown that *Ae. aegypti* had higher pyrethroid resistance than *Ae. albopictus*, indicating pyrethroid resistance was affected by ecological differences among the mosquitoes (Ping et al. 2001, Ponlawat et al. 2005, Jirakanjanakit et al. 2007, Pethuan et al. 2007).

2.3. Distribution Analysis of the Point Mutations in the VGSC (*kdr*) in *Aedes aegypti* in Vietnam

Two different mechanisms are involved in pyrethroid resistance: enhanced metabolic detoxification, and insensitivity of target sites. Pyrethroids target the voltage-gated sodium channel (VGSC) on the insects' neurons. Single amino acid substitutions in this channel are associated with pyrethroid resistance and are known as knockdown resistance (*kdr*) mutations. These *kdr*-type resistances have been observed in several mosquitoes, such as *Anopheles gambiae* Giles, *Anopheles stephensi* Liston, *Culex quinquefasciatus*, and *Ae. aegypti*. In *Culex* and *Anopheles*, *kdr* resistance is associated with the replacement of a leucine at position 1014 in segment 6 of domain II with either a phenylalanine or a serine (Martinez-Torres et al. 1998, 1999, Enayati et al. 2003). Brengues et al. (2003) described several mutations in a part of segment 6 of domain II in *Ae. aegypti*, and Saavedra-Rodriguez et al. (2007) found additional mutations at the same position (I1011M, I1011V, V1016G, and V1016I).

Recently, Yanola et al. (2008, 2011) found a novel mutation, F1534C, in segment 6 of domain III in DDT/permethrin-resistant *Ae. aegypti*. F1534C mutations seem to be distributed worldwide (i.e., South Asian, South East Asian, South and Central American countries, and Macaronesian islands). After the first description of the F1534C point mutation in *Ae. aegypti* collected from Thailand (Yanola et al. 2008, 2011), the same mutation was reported in succession in Vietnam (Kawada et al. 2009b), Grand Cayman Island (Harris et al. 2010), Madeira Island (Seixas et al. 2013), Brazil (Linss et al. 2014), Myanmar (Kawada et al. 2014), Venezuela (Alvarez et al. 2015), India (Kushwah et al. 2015), and Malaysia (Ishak et al. 2015). The mutations at 1016V were also reported worldwide. Two different types of the mutation at the same locus are distributed independently: valine to glycine replacements (V1016G) are commonly distributed in South East Asia (Rajatileka et al. 2008, Stenhouse et al. 2013, Kawada et al. 2014, Ishak et al. 2015, Pang et al. 2015), whereas valine to isoleucine replacements (V1016I) are common in South and Central America (Saavedra-Rodriguez et al. 2007, Martins et al. 2009, Siller et al. 2011, Marcombe et al. 2012, Linss et al. 2014, Alvarez et al. 2015).

No such genetic analyses of point mutations in the VGSC had been performed on *Ae. aegypti* in Vietnam, except on one colony (Long Hoa strain) (Enayati et al. 2003), until our nationwide survey (Kawada et al. 2009b). The objective of our survey was to elucidate the mechanisms of pyrethroid resistance by analyzing the presence of mutations in *Ae. aegypti* collected from used tires in Vietnam (Kawada et al. 2009a). Among the 706 larvae collected from 69 locations in Vietnam, no I1011M or I1011V mutations were found. Similarly, among the 735 larvae collected from 70 locations, not a single L1014F mutation was found. Only two V1016G heterozygous individuals were found at one location in the Binh Son district, Quang Ngai province, central east coast of Vietnam, from among 756 total larvae collected from 70 locations. The overall allelic frequency of V1016G was 0.26%. Conversely, the F1534C mutations on domain III were found to be distributed widely and had a high allelic frequency of 35.5% (269 individuals among 757 larvae). The same mutation was found at 53 collection sites among the 70 locations surveyed. The F1534C frequencies were lower in the middle and north part of Vietnam, but they were higher in the areas neighboring large cities such as Dong Ha, Hue, Da Nang, Tam Ky, Quang Ngai, Quy Nhon, and Nha Trang (Figure 2). The overall allelic frequency of F1534C was 21.6%, and the highest frequency was 87.5% in the collection site in Da Nang city. The percentage of homozygous individuals was also highest at this site (75.0%). Univariate analysis was performed by categorizing two parameters into each of two levels: >10 or <10 for the susceptibility indices, and >0 or 0 for the allele frequency or percentage of homozygous F1534C individuals. There was, however, no significant correlation between the susceptibility indices and either the allelic frequencies (χ^2 = 3.31, P = 0.069) or the percentage of homozygous F1534C individuals (χ^2 = 2.34, P = 0.126). The frequencies of the F1534C mutation in the southern mountainous area were exceptionally low, indicating the possible existence of another resistance mechanism, such as the enhanced detoxification pathway associated with pyrethroid insensitivity. The presence or co-presence of the other abovementioned resistance mechanisms might be one of the reasons for the lack of any significant correlation between the susceptibility indices and the F1534C frequencies. The overall percentage of homozygous F1534C individuals still seems to be low (7.4%) in the present situation. Extensive, frequent, and uncontrolled use of photo-stable pyrethroids, however, may set up a strong selection pressure in favor of this mutation and cause serious problems in controlling dengue fever in Vietnam.

The extensive use of photo-stable pyrethroids has been very common in southern Vietnam. Indoor residual spraying with DDT for malaria control had also been common in this area before the abandonment of DDT sprays in 1995. In Vietnam, 24 metric tons of DDT were used for residual treatment against malaria vectors in 1993 and 1994. Since the abandonment of DDT sprays, only pyrethroids (residual spraying with λ-cyhalothrin and α-cypermethrin and occasionally deltamethrin, and permethrin-impregnated bed nets) have been used in large amounts, unlike in other Asian countries (Nam et al. 2005, Zaim and Jambulingam 2007). The extensive use of DDT for IRS and the use of photo-stable pyrethroids, therefore, seem to have been very common in southern Vietnam. Both DDT and pyrethroid application for malaria vector control appear to have been intensively carried out in the interior and along the periphery of human habitation areas, where, incidentally, the breeding and resting sites of *Ae. aegypti* are located. This may account for the strong selection pressure for pesticide resistance in *Ae. aegypti*. As discussed below, selection for the F1534C point mutation may have taken place as a result of the extensive usage of DDT. DDT resistance, as well as the F1534C point mutation, may have been maintained in *Ae. aegypti* populations by further selection pressures from pyrethroid insecticides, such as permethrin, as both insecticides have the same target site.

3. DDT and Pyrethroid Resistance in *Aedes aegypti* in Yangon City, Myanmar

The first outbreak of dengue hemorrhagic fever was recorded in Myanmar in 1969. Since then, DHF has spread to 12 out of 14 states. During 1970–1991, 83,381 cases causing 3,243 deaths were recorded (Aung et al. 1996). The principal vector of DHF is *Aedes aegypti*, with *Ae. albopictus* acting as a secondary vector in suburban and rural areas. The growth of economic activities and traditional water storage practices increased the availability of vector breeding sites, thereby increasing the incidence of DHF cases from 5,621 in 2005 to 11,049 in 2006 (Oo et al. 2011). An *Aedes* mosquito control unit was established in 1968, and several feasibility studies aimed at the development of *Aedes aegypti* control methodology—including insecticide treatment, biological control by using larvivorous fish, environmental improvement, and health education—were undertaken (Aung et al. 1996). However, information regarding insecticide usage for DHF vector control in Myanmar is lacking, and few studies of insecticide resistance in DHF vectors have been published. During 1988–2005, 510 metric tons of organochlorine insecticides, 360 kg of organophosphate insecticides, and 8,157 kg of pyrethroid insecticides were used for malaria control in Myanmar. By contrast, only 1.2 metric tons of organophosphate insecticides were used for DFH control during the same period. These data indicate that large amounts of organochlorine insecticides, including DDT, were used in Myanmar until the use of these insecticides was banned in 2003. Thaung et al. (1975) reported that DDT was routinely used for control of the rat flea, *Xenopsylla cheopsis* (Rothschild), and that periodical monitoring of insecticide susceptibility in Yangon City indicated DDT resistance in rat fleas. The authors further suggested the presence of DDT resistance in *Aedes aegypti*, and the lack of resistance to other types of insecticides.

3.1. Insecticide Susceptibility of *Aedes aegypti* in Yangon City

High pyrethroid resistance in *Ae. aegypti* was observed in 2013 through larval collection and the simplified knockdown test (see section ***2-2***) (Kawada et al. 2014). With the exception of a single location in North Dagon where the susceptibility index was only 30, the maximum susceptibility index (36) was recorded for *Ae. aegypti* larvae in 54 of the 55 sites where this species was collected. The percentages of knocked down larvae at 0.4 ppm of *d*-allethrin, determined 30 min after treatment, were 0% for 42 collection sites and 10–30% for the remaining 12 collection sites. The susceptibility index of *Ae. albopictus* collected at a single site in the Dagon area was not high (18) . (Figure 3-A, B, Kawada et al. 2014). High resistance to DDT (less than 5% mortality at 7 townships by WHO tube test), permethrin (less than 10% mortality at 5 townships), and deltamethrin (less than 80% mortality at 3 townships) in the 7 townships were also detected in adult *Aedes aegypti* (Oo et al. unpublished data).

3.2. Co-occurrence of Triple Mutations in the VGSC in *Aedes aegypti* in Yangon City

No mutations at I1011 and L1014 were detected among 281 and 271 larvae sequenced. In contrast, 230 homozygous and 21 heterozygous 1016V (V1016G) individuals were detected among 285 larvae sequenced. The overall frequency and homozygous percentage of the V1016G point mutation were 84.4% and 80.7%, respectively (Figure 3-C). Additionally, 140 homozygous and 64 heterozygous 989S (S989P) individuals were detected among 218 larvae sequenced (frequency and homozygous percentage of the S989P mutation were 78.8% and 64.2%, respectively). Furthermore, 16 homozygous and 105 heterozygous 1534F (F1534C) individuals were detected among 323 larvae sequenced (frequency and homozygous percentage of the F1534C mutation were 21.2% and 5.0%, respectively) (Figure 3-D). Although the number of sampled individuals was small, no point mutations were detected in 20 larvae of *Ae. albopictus*.

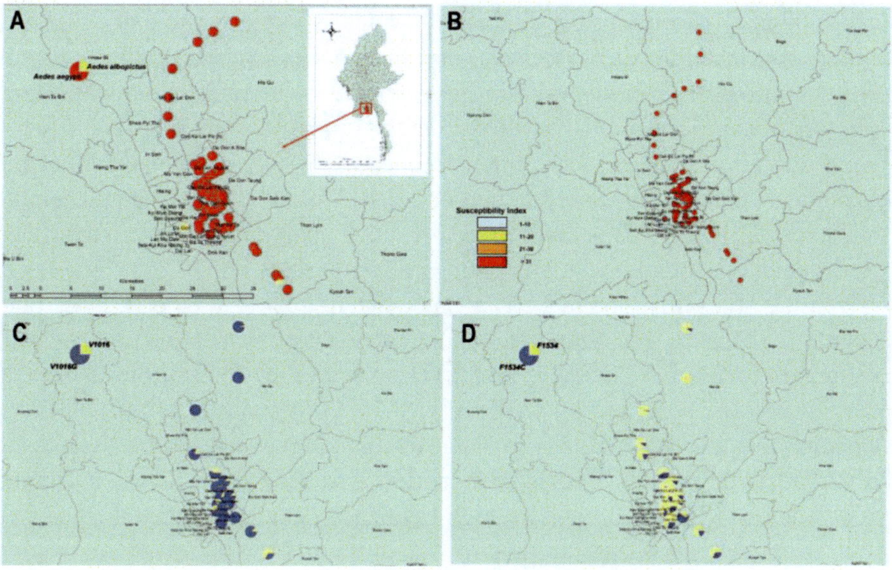

Each circle indicates the allelic composition of the point mutations/wild types.

Figure 3. Distribution of *Aedes aegypti* and *Ae. albopictus* larvae collected in used tires in Yangon City, Myanmar (A), distribution of susceptibility indices in *Ae. aegypti* larvae by the simplified knockdown bioassay (B), and distribution of the V1016G (C) and F1534C (D) point mutations in detected in *Ae. aegypti* larvae (Kawada et al. 2014).

Interestingly, three patterns of co-occurrence of point mutations were detected: V1016G/F1534C, V1016G/S989P, and V1016G/F1534C/S989P. The percentage of larvae homozygous for the V1016G mutation and the wild type 1534F was significantly higher (59.3%) than the percentage of larvae homozygous for the V1016G and F1534C mutations. By contrast, the percentage of individuals homozygous for both the V1016G and S989P mutations was notably high (65.7%). Additionally, co-occurrences of the above homozygous condition for three point mutations (V1016G/F1534C/S989P) were detected in two individuals among 204 individuals analyzed (0.98%) (Figure 4, Kawada et al. 2014).

High frequencies of the V1016G mutation in *Aedes aegypti*, namely, 3.1–61.8% (Stenhouse et al. 2013) and 23% (Rajatileka et al. 2008), have been reported in Thailand. The V1016G mutation has also been found in Vietnam, but its distribution appears to be limited and the frequency of the point mutation was relatively low (2 heterozygotes were found among 10 larvae collected at a single site in Quang Ngai Province) (Kawada et al. 2009b). At present, no other reports of this point mutation in other Southeast Asian countries have been published. It is interesting to speculate why the V1016G point mutation occurs with high frequency in Myanmar and Thailand. Myanmar shares the longest border (2,107 km) with Thailand among neighboring countries, and therefore the two countries may share environmental, sociological, and biological similarities.

In the study in Yabgon City, a moderate frequency of the F1534C point mutation (21.2%) was found; this frequency was comparable to the levels recorded in Vietnam (average, 21.6%) (Kawada et al. 2009b), but lower than the frequencies recorded in Thailand (59–79%) (Yanola et al. 2011), Grand Cayman Island (68%) (Harris et al. 2010), and Venezuela (100%) (Seixas et al. 2013). Additionally, the same mutation was recently reported in another DHF vector, *Aedes albopictus*, at high frequency (73%) (Kasai et al. 2011).

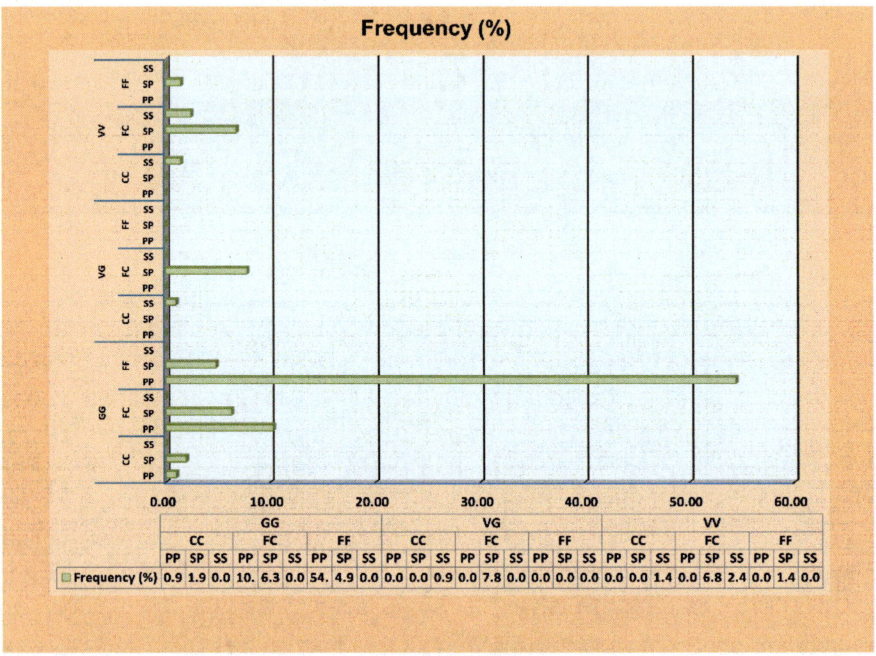

	GG									VG									VV								
	CC			FC			FF			CC			FC			FF			CC			FC			FF		
	PP	SP	SS	PP	SP	SS	PP	SP	SS	PP	SP	SS	PP	SP	SS	PP	SP	SS	PP	SP	SS	PP	SP	SS	PP	SP	SS
Frequency (%)	0.9	1.9	0.0	10.	6.3	0.0	54.	4.9	0.0	0.0	0.0	0.9	0.0	7.8	0.0	0.0	0.0	0.0	0.0	0.0	1.4	0.0	6.8	2.4	0.0	1.4	0.0

The total number of larval samples used for the analysis was 204.

Figure 4. Genotyping frequency of the V1016, F1534, and S989 point mutations in *Aedes aegypti* larvae collected from used tires in Yangon City.

The S989P mutation has always been linked to the V1016G mutation, although V1016G has sometimes been found in the absence of S989P (Srisawat et al. 2010). The observed co-occurrence of the V1016G and S989P mutations at high frequency in Myanmar corresponds to the former scenario. Srisawat et al. (2010) indicated that the co-occurrence of V1016G and S989P enhanced the resistance of *Aedes aegypti* to deltamethrin. However, those authors were unable to determine the effect of the S989P mutation alone on pyrethroid sensitivity, and were therefore unable to provide direct evidence of how the two mutations were together associated with resistance. Recently, Du et al. (2013) reported that the co-occurrence of the V1016G and S989P mutations was not additive or synergistic; however, the S989P mutation was concurrent with the V1016G mutation in a pyrethroid-resistant population of *Ae. aegypti*. Further studies regarding the synergistic role of S989P on V1016G are required.

Interestingly, Stenhouse et al. (2013) reported that in an *Aedes aegypti* population in Thailand, wild type 1016V mosquitoes were frequently homozygous for the F1534C mutation (43.5%); further, all heterozygous 1016 mosquitoes were also 1534 heterozygous (38.8%), and all homozygous V1016G mosquitoes (17.6%) were also homozygous for the wild type 1534F. These findings indicate that the wild type 1016V + homozygous F1534C, the homozygous V1016G, and the wild type 1534F, exist at equal frequencies. In contrast, in *Ae. aegypti* populations in Yangon City, predominance of the homozygous V1016G + homozygous wild type 1534F (59.3%) was detected; further, the percentage of mosquitoes possessing the wild type 1016V that were also F1534C homozygous was low (1.5%). Additionally, a small number of mosquitoes possessing the homozygous V1016G + F1534C (2.9%) and V1016G + F1534C + S989P (0.98%) combinations were detected. Recently, Linss et al. (2014) detected the co-occurrence of the V1016I and F1534C point mutations in *Aedes aegypti* populations in Brazil, and reported the widespread distribution of the wild type 1016V + homozygous F1534C and the homozygous V1016I + F1534C combinations in the country during the previous decade. Interestingly, the frequency of mosquitoes with the homozygous V1016I + F1534C combination in Brazil (average 20.5%) was higher than the reported frequencies of mosquitoes with the homozygous V1016G + F1534C combination. In Thailand, the F1534C mutation is considerably more common than the V1016G mutation (Yanola et al. 2011, Stenhouse et al. 2013).

Mosquitoes homozygous for the F1534C mutation are thought to be susceptible to deltamethrin (Stenhouse et al. 2013). The I1011M and F1534C mutations seem to play a role in decreasing susceptibility to Type I pyrethroids, such as permethrin; on the other hand, the L1014F (or S) and V1016G mutations play important roles in the insensitivity of the VGSC to Type II pyrethroids, such as deltamethrin (Hu et al. 2011, Du et al. 2013). Du et al. (2013) indicated that the contribution of the V1016I mutation to pyrethroid resistance was lower than that of the V1016G mutation. Therefore, in the *Aedes aegypti* population in Yangon City, the higher frequency of occurrence of the V1016G mutation in comparison to the F1534C mutation may indicate a history of pyrethroid treatment during previous vector control programs. First, the use of DDT or Type I pyrethroids might select for the F1534C mutations, with V1016G being subsequently selected for through the use of Type II pyrethroids. Hirata et al. (2014) expressed several types of VGSCs in *Xenopus* oocytes and found that the S989P+V1016G and F1534C haplotypes reduced channel sensitivity to permethrin by 100- and 25-fold, respectively, while S989P+V1016G+F1534C triple mutations reduced channel sensitivity to permethrin by 1,100-fold. S989P+V1016G and F1534C haplotypes reduced channel sensitivity to deltamethrin by 10-fold and 1-fold (no reduction), respectively, while S989P+V1016G+F1534C triple mutations reduced channel sensitivity to deltamethrin by 90-fold. The possible increase in homozygous V1016G and F1534C individuals in association with the S989P point mutation under increasing pyrethroid selection pressure may represent the worst case scenario for vector control programs.

In Myanmar, pyrethroid insecticides were first used for malaria control in 1992, and have since been constantly used (>500 kg annually). This intensive usage may explain the strong selection pressure toward resistance in *Aedes aegypti*, as previously described. Extensive use of DDT for malaria control before the use of this chemical was banned may also explain the development of pyrethroid resistance in that species.

4. Discovery of Point Mutations in the VGSC from African *Aedes aegypti* Populations, and Potential Phylogenetic Reasons for Gene Introgression

Aedes aegypti is found throughout West Africa from sea-level to at least 1,220 m in Nigeria, and from the coastal swamp zone to the northern Guinea savannas. Transportation and urbanization of new areas are major causes of the spread of *Ae. aegypti* (Surtees 1967). Yellow fever is endemic in Ghana, and major outbreaks involving 319 cases and 79 deaths occurred in 1969–1970 in the northern part of the country. In December 2011, the Ministry of Health of Ghana declared a yellow fever outbreak when cases were recorded in three districts located in the midwestern part of the country. A total of three laboratory-confirmed cases and seven deaths were reported (DREF final report "Ghana: Yellow fever outbreak," http://adore.ifrc.org/ Download.aspx? FileId=28108). *Ae. aegypti* is one of the most important yellow fever vectors implicated in the Ghana outbreaks (Agadzi et al. 1984). Although there have been no reports of dengue fever outbreaks in Ghana, it has been detected in the adjacent countries of Côte d'Ivoire and Burkina Faso, both of which share borders with Ghana. Increasing migration of people across the borders of these countries and the absence of organized mosquito control in Ghana may lead to dengue fever transmission in Ghana in the future (Appawu et al. 2006). A recent seroprevalence survey in Ghana revealed the presence of IgM and IgG dengue antibodies in 3.2% and 21.6%, respectively, of children with confirmed malaria. This indicates the possible co-infection of dengue fever and malaria, and previous exposure of the children to dengue virus (Stoler et al. 2015).

Mapping the nation-wide distribution of insecticide resistance in vector mosquitoes in Ghana will provide beneficial information for forecasting dengue and yellow fever outbreaks and developing effective control measures. Differences in insecticide susceptibilities associated with seasonal or regional differences in the distribution of the subspecies *Ae. aegypti aegypti* (Aaa) and *Ae. aegypti formosus* (Aaf) are also of interest. The Aaf subspecies, which originated in African forest areas, is believed to be the ancestor of *Ae. aegypti* sensu lato. The Aaa subspecies is predominantly anthropophilic and adapted to the human environment, while Aaf is more associated with forest environments (Mattingly 1957). Adults of Aaa prefer indoor environments and use artificial water containers for oviposition, while Aaf prefer outdoor environments and forest edges and breed in natural containers such as tree holes, rock pools, and plant axils. Aaa is highly susceptible to dengue and yellow fever virus, and is considered to be a more efficient virus vector than Aaf (Failloux et al. 2002).

4.1. Insecticide Susceptibility of Ghanaian *Aedes aegypti* Colonies

In the first collection in November and December 2013, Aaa appeared dominant in Accra (82.7%), Kintampo (87.0%), and Tamale (79.3%), whereas its prevalence was somewhat lower in Kumasi (65.2%). Conversely, in the second collection performed in September 2014, the prevalence of Aaa in Accra (46.4%), Abuakwa (25.5%), and Kintampo (60.0%) was lower than in the 2013 collection, while the rate in Kumasi was in the same range as in 2013 (63.7%) (Figure 5). The rainy season in Ghana starts in March and lasts until the end of October. The collection dates therefore corresponded to the beginning of the dry season in 2013 and the late rainy season in 2014. Accordingly, the proportion of Aaa collected in samples from used tires in Accra and Kintampo was higher in the dry than the rainy season, although no such difference was observed in Kumasi. It is interesting that there was no such seasonal difference in subspecies composition in Kumasi. This was apparently because of the relatively consistent precipitation throughout the year in Kumasi (monthly average precipitation in 2012 and 2013 was 162.4 mm in the rainy season and 88.9 mm in the dry season) as compared to Accra (68.2 mm and 29.6 mm for the wet and dry seasons, respectively) and Kintampo (75.1 mm and 46.6 mm for the wet and dry seasons, respectively, in adjacent Tamale) (Kawada et al. 2016).

Since there were no differences in the susceptibility to permethrin and DDT between Aaa and Aaf, susceptibilities to the insecticides were compared using mixed colonies of both subspecies. High resistance to DDT (less than 70% mortality with 1 h contact) was observed in all mosquito colonies. Susceptibilities to permethrin were relatively higher (0.75%) in all colonies as compared to those for DDT. Resistance to permethrin was, however, suspected in the Accra (<90% mortality) and Abuakwa colonies (81.3% mortality). For the Kumasi and Kintampo colonies collected in 2014, mortalities with permethrin were higher than in the other colonies, but were less than 100% (98.8% and 95.0% mortality, respectively), indicating the possible existence of resistance or tolerance to permethrin (Figure 5, Kawada et al. 2016).

The above results are perhaps the first to report the resistance of *Ae. aegypti* s. l. to DDT and permethrin in Ghana, although DDT resistance in *Ae. aegypti* is common in countries adjacent to Ghana, such as Côte d'Ivoire (1968), Togo (1969), and Benin (1968) (WHO 1986). High resistance to DDT seems to be widespread and resistance to pyrethroids is also suspected to be common in Ghana. Although the concentration of DDT and pyrethroids used and their method of application for the control of *Ae. aegypti* in Ghana is unknown, it is clear that these insecticides were one of the causative factors in the resistance of *Ae. aegypti* s. l. populations in Ghana. Organochlorine pesticides were popular and extensively used by farmers in Ghana, with lindane commonly used for pest control on cocoa, vegetables, and maize, and endosulfan on cotton, vegetables, and coffee. DDT and lindane were once employed to control ectoparasites of farm animals and pets in Ghana (Fianko et al. 2011). Lambda-cyhalothrin and cypermethrin are used by vegetable growers on tomato, pepper, okra, eggplant, cabbage, and lettuce farms. The contamination caused by the aforementioned pesticides to the breeding area may have acted as a selection pressure for the development of resistance in *Ae. aegypti* populations.

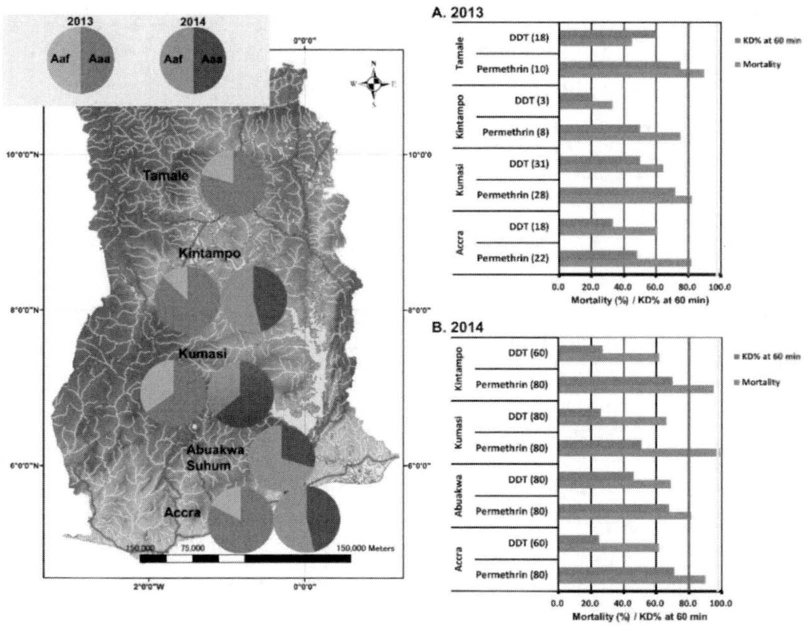

Shape file map (left map) available from DIVA-GIS (http://www.diva-gis.org/gdata) was used using ArcGIS 10.2 (ESRI Japan Corp., Tokyo, Japan). Figures in parenthesis (right graph) indicate the number of mosquitoes used for the test. Mosquito colonies mixed with Aaa and Aaf emerged from the field collected larvae were used for the test in the collection in 2013. The F1 adult mosquitoes emerged from the eggs collected from field collected larvae were used for the test in the collection in 2014. A mortality rate between 98% (yellow line) and 100% is considered to indicate susceptibility; 80 (red line) – 97% mortality suggests the possibility of resistance that needs to be confirmed. Mortality <80% indicates resistance by WHO criteria.

Figure 5. Subspecies composition of *Aedes aegypti aegypti* (Aaa) and *Ae. aegypti formosus* (Aaf) at 5 collection sites in Ghana and insecticide susceptibility of adult *Aedes aegypti* (mixed colonies of Aaa and Aaf) by the WHO tube test (Kawada et al. 2016).

The indirect effects of long lasting insecticidal nets (LLINs), IRS, and other insecticide treatments for malaria control have also contributed to the development of DDT and pyrethroid resistance in *Ae. aegypti* populations in Vietnam, as previously reported (Huber et al. 2003, Vu et al. 2004, Kawada et al. 2009a, 2009b). The malaria vector *Anopheles gambiae* s. l. has developed a high resistance to DDT and pyrethroid insecticides in an area of southwestern Ghana where the species was susceptible to these chemicals just a decade ago (Kristan et al. 2003, Kudom et al. 2012). The use of insecticides in LLINs and IRS in the Ghana National Malaria Control Program is believed to be the major cause for the cross-resistance between DDT and pyrethroids. This was mainly attributable to the *kdr* gene (Kudom et al. 2012), as reported in the adjacent countries of Mali (Fanello et al. 2003) and Burkina Faso (Diabaté et al. 2002). Extensive use of DDT for malaria control before it was banned and of pyrethroid treatment for malaria vector control may have exerted a strong selection pressure on *Ae. aegypti* (especially Aaa).

4.2. Point Mutations in the VGSC Found in Ghanaian *Ae. aegypti*

No mutations at S989, I1011, or L1014 were detected among 707, 756, and 734 mosquitoes sequenced, respectively. Conversely, F1534C mutations were detected at high frequency: 294 homozygous and 259 heterozygous individuals were found among 759 mosquitoes sequenced. The allelic frequencies of F1534C were higher in Accra (68.4%), Kumasi (64.6%), and Kintampo (58.3%) than in other places. The allelic frequencies of F1534C in Aaf were also higher in Accra (52.6%), Kumasi (60.0%), and Kintampo (45.2%) than in other places. The frequency of the F1534C allele and the ratio of F1534C homozygous mosquitoes in Aaa was significantly higher than in Aaf in the mixed populations from all collection localities in 2013 and 2014 (Figure 6). Additionally, one heterozygous point mutation (V1016I) was found in Accra among 732 mosquitoes. Homozygous F1534C was concurrently found in this individual (Aaa). No V1016I mutations were found in Aaf mosquitoes.

The above study provided the first description of F1534C and V1016I mutations in African *Ae. aegypti* s. l. populations. Accra and Kumasi are the two largest cities in Ghana, each home to one to two million people. Tamale is the third largest city with a population of approximately 400,000 people. Kintampo and Abuakwa, both of which have populations of approximately

40,000, are much less populated compared to the other sampled cities. Allelic frequencies of F1534C and the percentage of individuals homozygous for the same point mutation were higher in the two large cities, Accra and Kumasi, than in the other collection locations. Interestingly, both the allelic frequencies and homozygous percentages of F1534C were significantly higher in Aaa than in Aaf, although we did not detect a significant difference in susceptibility to DDT and permethrin between the two subspecies. The above discrepancy might suggest that the F1534C mutation is not a single resistance mechanism but is combined with other unknown mechanisms such as metabolic factors. Recently, some studies reported the role of glutathione-S-transferases (GST) in the cross-resistance between DDT and pyrethroids in mosquitoes (Lumjuan et al. 2011, Riveron et al. 2014). Riveron et al. (2014) demonstrated that a single amino acid change in the GST gene (L119F) confers a high level of metabolic resistance to DDT in *Anopheles funestus*. The authors also showed that this mutation was strongly related to the metabolism of permethrin. Several Epsilon GST genes were reported to play a role in pyrethroid resistance in *Ae. aegypti* (Lumjuan et al. 2011). The above new findings, as well as other metabolic factors, should be taken into consideration for further study.

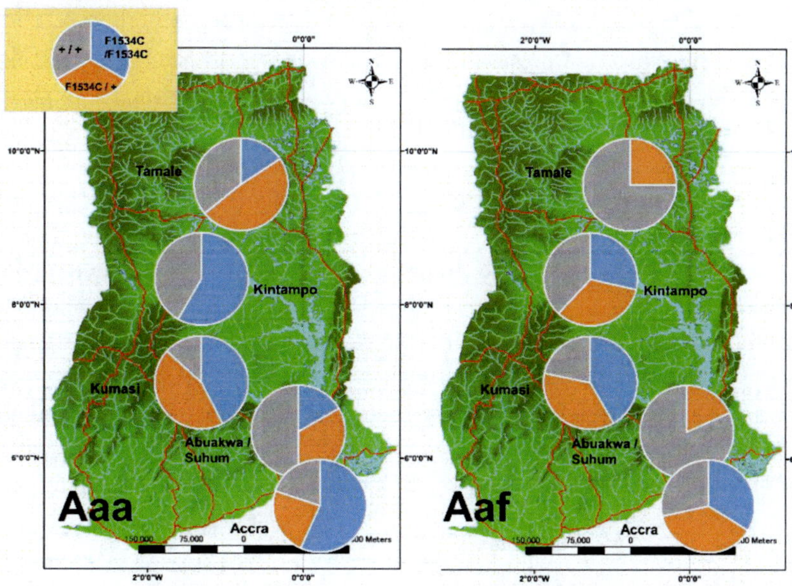

Figure 6. Allelic frequency of F1534C in *Aedes aegypti aegypti* (Aaa) and *Ae. aegypti formosus* (Aaf) collected in Ghana (Accra, Abuakwa/Suhumu, Kumasi, Kintampo, and Tamale) (Kawada et al. 2016).

4.3. Phylogenetic Analysis of Ghanaian *Aedes aegypti* Based on the Intron Located in the VGSC Exon

Two types of introns between exon 20 and 21 in Ghanaian *Aedes aegypti* populations were detected in the above study, Ghana 001 (250 bp) and Ghana 257 (234 bp). The point mutations at 1534F (F1534C) on exon 31 were found to be strongly linked with the intron of the former group. When the two intron types were treated as two alleles, strong linkage disequilibrium was observed between the mutation at 1534F and the two intron types. The phylogenetic tree showed an interesting topology (Figure 7). Two large clades were observed: Clade 1 consisted of the sequences from southeastern Asia and South and Central America, with two from Kenya and one from Ghana, while Clade 2 consisted of the remaining African samples and strongly supported Asian and American branches. The Ghana 001 intron shared the same sequence with most of the other Asian and South American sequences within Clade 1. Most African sequences were placed in Clade 2, and the Asian and American sequences were distributed in three strongly supported monophyletic clades within Clade 2.

Martins et al. (2009) reported two haplotypes, group A (250 pb) and group B (234 pb), in the intron between exons 20 and 21 on domain II of the VGSC, and these haplotypes had pronounced differences in both sequence and size in Brazilian *Ae. aegypti*. The authors also noted that point mutations at 1011I (I1011M) on exon 20 appeared in half of the group A sequences, whereas no such mutation occurred in the group B sequences. The same kind of evidence of linkage equilibrium was reported by Saavedra-Rodriguez et al. (2007). The authors found the same intron as reported above (group A) strongly linked with the V1016I mutation and hypothesized that a genetic sweep of the V1016I allele and its proximate intron sequences has occurred through DDT-driven and subsequent pyrethroid-driven selection. In the study of Ghanaian *Ae. aegypti* populations, the same two intron types, Ghana 001 for group A and Ghana 257 for group B, were detected. Interestingly, the point mutations at 1534F (F1534C) on exon 31 were found to be strongly linked with the intron of group A. Furthermore, phylogenetic analysis using this intron in the present study clearly showed that the two Ghanaian haplotypes belonged to two haplotype groups (Clades 1 and 2). Given *Ae. aegypti* was originally from Africa, Clade 2 is thought to be the ancestral clade of Clade 1 because Clade 2 contained most of the African haplotypes. Interestingly, one of the two Ghanaian haplotypes (Ghana 001) and two Kenyan haplotypes were placed in

Clade 1, suggesting that those haplotypes were introduced from other continents, such as Asia or South or Central America.

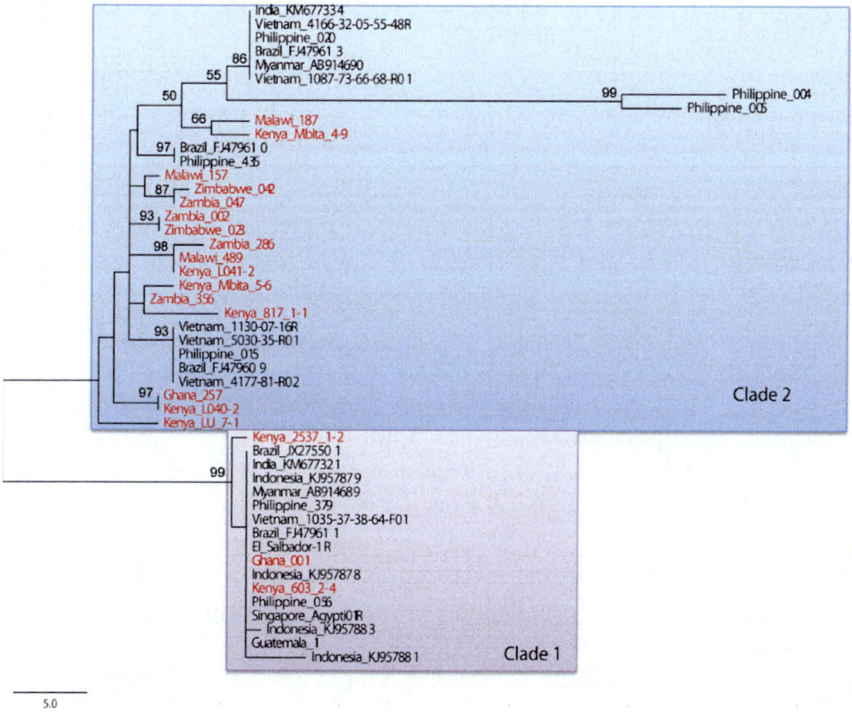

The evolutionary history was inferred using the Maximum Parsimony method using MEGA6. Tree No. 1 out of the four most parsimonious trees (length = 116) is shown. The percentages of replicate trees in which the associated sequences clustered together in the bootstrap test (1,000 replicates) are shown next to the branches51. The MP tree was obtained using the Subtree-Pruning-Regrafting (SPR) algorithm (pg. 126 in tha manual of PHYLIP version 3.69) with search level 1 in which the initial trees were obtained by the random addition of sequences (10 replicates). The tree is drawn to scale, with branch lengths calculated using the average pathway method (see pg. 132 in the manual of PHYLIP version 3.69) and are in the units of the number of changes over the whole sequence. The analysis involved 48 nucleotide sequences. All positions containing gaps and missing data were eliminated. There were a total of 209 positions in the final dataset.

Figure 7. Maximum Parsimony analysis of *Aedes aegypti* (Kawada et al. 2016).

Aedes aegypti is thought to have originated on the African continent. The sub-Saharan part of the continent still contains Aaf, which is believed to be the ancestor of *Ae. aegypti* s. l. The subspecies that has become domesticated and adapted to anthropogenic environments (Aaa) expanded its habitat around human domiciles and has been dispersed by human movement. *Ae. aegypti aegypti* spread to the western hemisphere in the 17th century, to the Mediterranean coastal area in the 18th century, and to tropical Asian and Pacific islands in the 19th and 20th centuries. This subspecies was eradicated from the Mediterranean area in 1950s and from South America between 1950 and 1960. However, it has reinfested most of the countries from which it was eradicated (Rodhain and Rosen 1997). The *Ae. aegypti* eradication program, initiated by the Pan American Health Organization (PAHO) in the 1940s and 1950s to prevent urban epidemics of yellow fever, was successful in most of the countries in South and Central America, resulting in a dramatic decrease in the distribution of mosquito populations. However, the discontinuation of the eradication program in 1970s led to reinfestation, and by 1995 *Ae. aegypti* regained a similar distribution to what it had in the 1940s (Gubler 1997). The worldwide eradication programs, presumably with organochlorine insecticides, in 1940s and 1950s may have caused high selection pressure on the mosquito populations and expanded the distribution of insecticide-resistant *Ae. aegypti* populations (WHO 1986). Selection of the F1534C point mutation could be hypothesized to have taken place during this period. DDT resistance in *Ae. aegypti* was first reported in the Caribbean countries in the 1950s and resistance persists in almost all regions that had achieved *Ae. aegypti* eradication, despite the fact that DDT is no longer used. DDT resistance, as well as the F1534C point mutation, may have been maintained in *Ae. aegypti* populations by selection pressures from pyrethroid insecticides, such as permethrin, as both insecticides have the same target site.

Used and discarded tires are one of the most important breeding sites for *Ae. aegypti*. They provide a habitat for the larvae and are capable of supporting larval development immediately after they are discarded. It is noteworthy that there has been worldwide focus for the past three decades on the dispersal of shipping containers of used tires that contain breeding mosquitoes (Reiter and Sprenger 1987). The species that has played the leading part in this event is *Ae. albopictus*, which together with four other Japanese mosquito species, is thought to have arrived at the western coastal ports of the United States in used tires by 1983, becoming widely distributed in the United States and Brazil by 1986. Since 1986, tire shipments infested with *Ae. albopictus* have been found in South and Central America, South and

West Africa, Oceania, and European countries (Reiter 1998). Used tire trading occurs worldwide and involves complicated commercial networks in many countries, including South Africa, Kenya, Uganda, Niger, Nigeria, and Ghana (Reiter and Sprenger 1987, Reiter 1998). The history of the worldwide *Ae. aegypti* invasion in association with human activity may be longer than that of *Ae. albopictus*.

The worldwide transportation of vector mosquitoes in accordance with human activity such as the trading of used tires may have resulted in the introgression of Asian or Central and South American haplotypes into Ghanaian *Ae. aegypti* populations. It is not known whether Ghanaian F1534C mutations "hitched a ride" with the above haplotypes or were selected convergently after the above introgression. However, strong linkage disequilibrium between the F1534C mutation and intron haplotypes may support the introgression hypothesis. Discovery of the V1016I mutation, although it was only found in one heterozygote, and the co-occurrence of this mutation with F1534C, may be noteworthy. The high frequency of F1534C and the co-occurrence of V1016I with this mutation may therefore explain one of the possible histories of introduction of these mutations into the Ghanaian *Ae. aegypti* population from South and Central America. The importance of the sequential evolution of F1534C and V1016I was suggested in Mexican *Ae. aegypti* populations by Vera-Maloof et al. (2015). The V1016I mutation was unlikely to have evolved independently because of low fitness, while the F1534C mutation evolved first but conferred only low level resistance. The V1016I mutation then rapidly evolved from the 1016V/F1534C haplotype under the high pressure of pyrethroids since these double mutants confer higher pyrethroid resistance. These authors suggested that knowledge of the frequencies of mutations in S6 domains II and III is important for predicting the potential of a population to evolve *kdr*. They also sounded an alarm that susceptible populations with high 1016V/F1534C frequencies are at high risk for *kdr* evolution. Great attention to the genomic status of Ghanaian *Ae. aegypti* populations should therefore be required for predicting the evolution of pyrethroid resistance.

5. DDT and History of Malaria Control in Japan

Japanese people had been suffering from malaria from the ancient times and the historical facts that the Emperor Horikawa (1078-1107), Taira-no-Kiyomori (1118-1181) Ikkyu Sojun (1394-1481), etc. suffered from the *Plasmodium vivax* (*Pv*) malaria are famous stories. In 1900s, more than 1,400 farmer soldiers (Tonden-Hei) among 8,200 people were reported to be infected with *Pv* malaria in Hokkaido. Of eight kinds of *Anopheles* mosquitoes recorded in Japan excluding Ryukyu archipelago, *Anopheles sinensis* Wiedemann and *Anopheles lesteri* Baisas et Hu were transmitting *Pv* malaria (Kurihara 2002). *Pv* malaria had been indigenous at Lake Biwa shore in Shiga Prefecture, Kitsu-gawa and Kamo-gawa river basin in Kyoto Prefecture, Tone-gawa river basin in Kanto area, Shinano-gawa and Agano-gawa river basin in Niigata Prefecture, Nagara-gawa and Ibi-gawa river and Kiso-gawa river basin in Gifu Prefecture, and Kuzuryu-gawa basin in Fukui Prefecture.

In 1940s, 6 million demobilization soldiers and immigrants including 600,000 malaria-infected people came back to Japan in short time after World War II from tropical area and brought back *Pv*, *Plasmodium falciparum* (*Pf*), and *Plasmodium malariae* (*Pm*) malaria ("imported malaria", *Pv:Pf:Pm* = 43:10:1) (Morishita 1963). However, in less than 5 years postwar under such severe conditions, both indigenous and imported malaria cases had disappeared in Japan (Ohtsuru 1958). The reason why such rapid disappearance resulted in is still unsettled. Sufficient medical treatment with anti-malaria drug (quinine) and improvement of self-protection, such as mosquito-coil and bed net usage, decrease in number (due to pesticide treatment in paddy field) and less anthropophyly and endophyly of vector mosquitoes (*An. sinensis* and possibly *An. lesteri*), and the political reduction of area and seasonally regulated use of rice field which served as breeding site for vector mosquitoes, etc. are reasonable answer for the above question (Ohtsuru 1958, Morishita 1963). Surprisingly, there was almost no area where active measure for malaria prevention, such as DDT indoor residual spray (IRS) was done in mainland of Japan.

Pv malaria was prevalent nationwide in Japan before World War II, but it gradually decreased and disappeared except for Shiga, Fukui, Ishikawa, Toyama, Niigata, Aichi, Tochigi, and Kochi prefectures (Sawada 1949). Hikone city (Shiga prefecture) was once malaria endemic area reporting > 20% of malaria patients. Primary 5 year-plan (1949-1953) and the secondary 5 year-plan (1954-1959) for malaria prevention were promoted under the direction of the occupation US army. First of all, the knowledge concerning malaria was propagated among school children by the leaders in the region. Schools, companies, factories, and other business establishments were established as bases for malaria control. Malaria Research Center promoted the environmental improvement and the use of anti-malaria drugs and insecticides, such as DDT EC (emulsifiable concentrate) and pyrethrum EC. As a result, no malaria case has been reported since 1954 in Hikone city.

The history of the malaria in Ryukyu archipelago takes on a different aspect from the above mainland case. Six species of *Anopheles* mosquitoes are recorded in Ryukyu archipelago, among them *Anopheles yaeyamaensis* Somboon & Harbach, sp. n. was important vector of *Pf* malaria. The US navy task force closed in upon 380 km eastern marine of Ishigaki Islands in October, 1944. Miyazaki regiment (No. 45 regiment of Japanese Independent Hybrid Army) took the intercept formation in Yaeyama archipelago. They used the natural caves as their positions. Although Japanese army had prepared the malaria prevention measures, 90% of the regiment's soldiers were infected with malaria (Haibara 2012). First air attack was done against Ishigaki island on Oct. 12, 1944. US task force landed on Okinawa island on Apr. 1, 1945 and repeated air attacks by thousands of carrier-borne aircrafts and bombardments from warships were done against Yaeyama archipelago. At 12 days after landing on Okinawa island (April 13, 1945) US army started air spraying of DDT by plane (Douglas C-47 Skytrain). They had 7 malaria investigation teams, 3 malaria control teams, and 3 public sanitation teams. They also performed the medical examination and blood check of civilian, and malaria case in this island rapidly decreased, while the other diseases such as filariasis rapidly increased. In Okinawa air spraying with DDT was common in 1946 - 1947 (this was actually larviciding !!) but this type of large indiscriminative air spraying and the abandonment and subsequent re-opening of numerous rice paddies led to upset in the balance of nature. By neglecting the preventive action on the ground, resurgence of malaria and Japanese encephalitis caused the deaths among US military government. Indiscriminating air spraying of DDT was then cancelled and they shifted to on-the-ground prevention and topical air spraying with small plane. The height and wind velocity at the air spraying with DDT was regulated (>150 feet and <12 mile/sec) in 1949. The air spraying was still common in 1940s and it was 1952 when full scale on-the-ground operation was started (Hosaka 1992).

Yaeyama archipelago still was the endemic area of malaria and the number of patients from 1954 to 1957 exceeded 3,000. US government dispatched a medical entomologist, Dr. C. M. Wheeler, as a leader of the new control plan targeting adult vectors having ability of transmission of malaria

(Wheeler Plan). He started 3 year-plan (Aug. 15, 1957) according to the measure of WHO control program that was IRS to the whole houses in all islands irrespective of malaria prevalent or not. He also made a guideline that primaquine was administered for 2 weeks after chloroquine administration. Accordingly, number of malaria patients dramatically decreased from 1,730 (1957) to 370 (1958), 58 (1959), 4 (1960), and zero in 1961 at last. IRS of DDT was terminated in 1962 and the US budget for malaria control was also terminated in 1963. The total budget for the program was $16,338,000 (Haibara 2012).

Via the abovementioned process with big effort which was made by the officials and the people in Japan incorporated with the occupation US army, malaria was expelled from Japan.

Tsukamoto et al. (1957) reported high DDT resistance in houseflies collected in Hikone city, while no such resistance was observed in *An. sinensis*, *Culex pipiens*, *Culex tritaeniorhynchus*, and *Aedes japonicus*. Umeda et al. (1990) suggested the presence of *kdr* type high pyrethroid resistance in *Culex tritaeniorhynchus* collected from paddy field in Chinen town (Yasutomi and Takahashi 1989), Okinawa. The authors could not find causative factor for the *kdr* resistance in this species, since "neither DDT nor pyrethroids had ever been used in paddy fields in Chinen town". However, it might be apparent that abovementioned DDT sky spray for malaria control in 1945-1962 had played an important role in developing pyrethroid resistance in this species. Toma et al. (1993) reported that *Culex tritaeniorhynchus* collected in the northern, central, and southern part of Okinawa in 1970-1972 showed resistance to DDT, while same species collected in the central part of Okinawa and Ishigaki island were susceptible to DDT. Gentry and Hubert (1957) reported DDT and other organochlorine insecticide resistance in *Culex quinquefasciatus* on Okinawa. Surprisingly, there is few report on the insecticide resistance of malaria vector in Ryukyu archipelago. Toma et al. (1998) reported high DDT and pyrethroid resistance in *An. sinensis* collected in Chinen town. The authors, unfortunately, could not perform the susceptibility test for *An. minimus* (*An. yaeyamaensis*) to the both insecticides in this report. All the abovementioned reports seem to suggest the causative role of DDT to pyrethroid resistance.

6. Pyrethroid and DDT Resistance Status of *Aedes albopictus* in Japan

Japan, which is located in a temperate area, experienced epidemic dengue outbreaks. *Aedes aegypti* was once inhabiting in Ogasawara and main Okinawa island in Japan, but this species is believed to become extinct in Japan at present. Temporal inhabitation of *Ae. aegypti* was recorded in 1944-1952 in Ushibuka town, Kumamoto prefecture, but no larva was observed in the survey in 1955 and 1957 (Hotta 1998). In 1942, one warship from Malay peninsula visited in Nagasaki port. Thirteen among 75 sailors on this ship who had been diagnosed as dengue fever disembarked on the port and had contacts with Japanese laborers and the sailors' family. This triggered the dengue outbreak in Japan. The dengue fever patients rapidly increased to 2,295 in one month and 13,323 in the next 2 months in Nagasaki city. This outbreak expanded to other cities, such as Sasebo, Hiroshima, Kure, Kobe, and Osaka within this year and the pandemic lasted until 1944, resulting in total number of 200,000 patients. This was one of the biggest pandemicity in the temperate zone. *Aedes albopictus* is assumed to be the vector responsible for the above dengue outbreaks in Japan (Hotta, 1998). *Ae. albopictus* inhabits tropical regions but is also found in temperate regions, including Japan, Europe, America, and Australia (Powers and Logue 2007, Bonilauri et al. 2008). *Ae. albopictus* is also the main vector of the chikungunya virus (CHIK) (Powers and Logue 2007, Bonilauri et al. 2008). Although DHF has not been prevalent in Japan for the past 50 years and CHIK outbreaks have never been reported in Japan, the possibility of viral disease outbreaks exists because of the widespread distribution of the vector *Ae. albopictus*. Although many imported cases of dengue fever have recently been reported (Takasaki 2011), no autochthonous cases had been confirmed in Japan for 70 years (Hotta 1998) until a patient with no history of overseas travel was first diagnosed in Tokyo on August 26, 2014, and a total of 160 autochthonous cases were subsequently confirmed (Kutsuna et al. 2015).

Only a few studies of insecticide resistance in *Ae. albopictus* have been conducted in Japan. Toma et al. (1992) reported on the susceptibility of seven colonies of *Ae. albopictus* from the Ryukyu Islands to 11 types of insecticides, including DDT, organophosphates, carbamates, and pyrethroids, and concluded that all strains were susceptible to all insecticides, except DDT. The dipping method was used by Suzuki and Mizutani (1962) to test the susceptibility of *Ae. albopictus* larvae collected from Kawasaki (Tokyo) and

Kawashima (Nagasaki) to organochlorines and organophosphates. The larvae showed tolerance to the above insecticides. Moreover, Kawada et al. (2010) found possible pyrethroid resistance in *Ae. albopictus* larvae collected from some locations in Nagasaki City. In this section, the pyrethroid resistance status of *Ae. albopictus* collected from Nagasaki City is discussed and compared to specimens collected from several other locations in Japan. The possible mechanism of resistance is also discussed.

6.1. Insecticide Susceptibility of *Aedes albopictus* Collected in Catch Basins in Nagasaki City, Japan

The susceptibility indices (see section ***2-2***) of *Aedes albopictus* collected from Nagasaki City and other places in Japan ranged from 12 to 36. Among the 20 colonies from Nagasaki City, 10 showed the highest susceptibility index (36), six colonies showed an index of 24–30, and four colonies showed an index of 12–18, indicating that more than half of the Nagasaki colonies were highly tolerant to *d*-T_{80}-allethrin. Among eight colonies collected from other places in Japan, two colonies showed a susceptibility index of 36, two colonies showed an index of 24–30, and the other four colonies showed an index of 12–18 (Figure 8, 9).

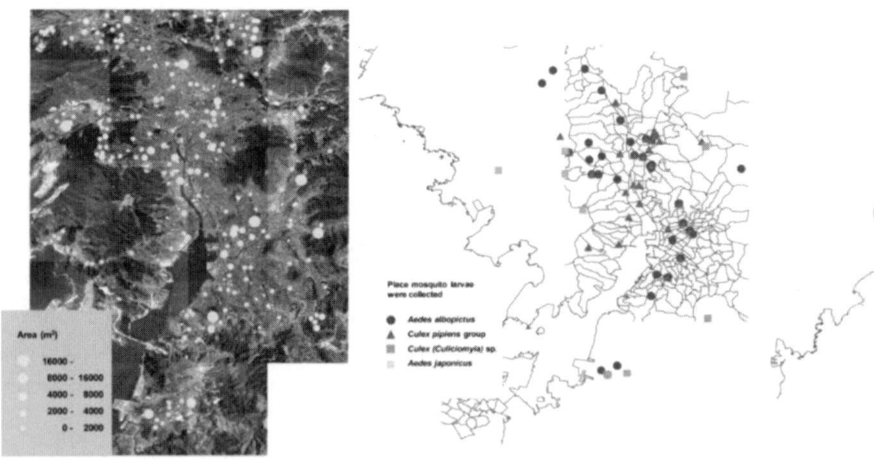

Figure 8. The location and area of the parks investigated in Nagasaki city (left map), and the location of the parks where mosquito larvae were collected (right map) (Kawada et al. 2010).

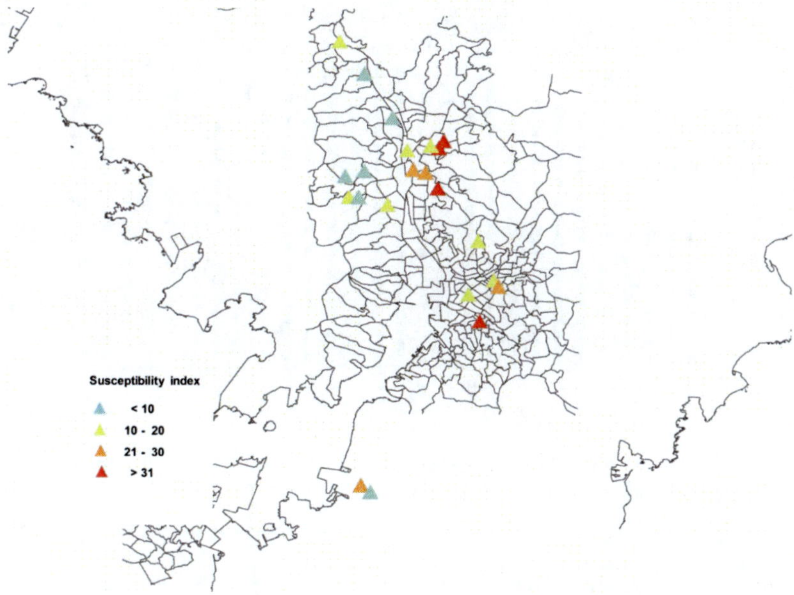

Figure 9. The location of the parks where the larval bioassay for *Aedes albopictus* was performed and the susceptibility index (see *2.2*) observed at each site in Nagasaki city (Kawada et al. 2013).

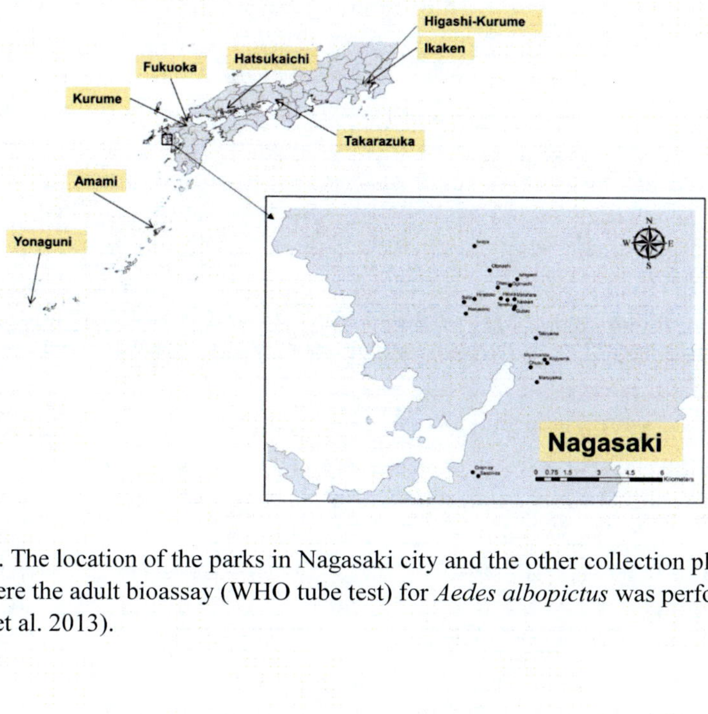

Figure 10. The location of the parks in Nagasaki city and the other collection place in Japan where the adult bioassay (WHO tube test) for *Aedes albopictus* was performed (Pujiyati et al. 2013).

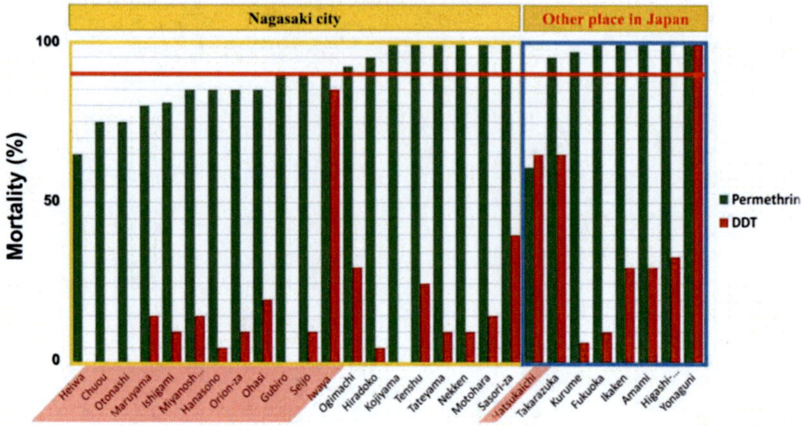

Twelve colonies collected in Nagasaki city and one colony collected in other places showed <90% mortality. All colonies except for Yonaguni colony showed high resistance to DDT.

Figure 11. Permethrin and DDT resistance status in *Aedes albopictus* collected in Nagasaki city and the other places in Japan by WHO tube test (Pujiyati et al. 2013).

Adult susceptibilities to permethrin and DDT of F1 colonies collected from Nagasaki City and several other parts of Japan (Figure 10) are compared by WHO tube test. When 90% mortality was adopted as the threshold for the detection of resistance, 12 of the 20 colonies collected from Nagasaki City were categorized as permethrin-resistant. On the other hand, only one colony (Fukuoka) was categorized as permethrin-resistant among the eight colonies collected from other places in Japan. Using the same criteria as above, all colonies, except the Yonaguni colony, were categorized as DDT-resistant. Nineteen Nagasaki colonies showed very low mortality (<40%), indicating that these colonies were highly resistant to DDT (Figure 11, Pujiyati et al. 2013).

The above study confirmed that most *Ae. albopictus* colonies collected in Nagasaki City, and one colony collected in Fukuoka City, showed resistance to pyrethroids. Moreover, it was surprising that almost all colonies collected in Japan, except for the Yonaguni colony, were found to be resistant to DDT. This might be the first report of widespread DDT resistance in this species in Japan, although some studies have also reported DDT resistance in some local Japanese populations (Suzuki and Mizutani 1962, Suzuki 1962, 1963, Miyagi et al. 1989, Toma et al. 1992). Widespread DDT resistance in mosquitoes and other insects in Japan are probably attributable to the widespread use of the chemical in the nationwide control programs of 1945–1962 (Toma et al. 1992, Kasai et al. 2007). It is noteworthy that the Yonaguni colony in particular was susceptible to DDT out of all the colonies collected in Japan. The same result was reported by Toma et al. (1992), in which the Yonaguni and Minami Daito colonies were susceptible to DDT from among several colonies collected from Okinawa Prefecture. The above facts confirm that extensive DDT treatment was not performed in those islands during the vector control program (Miyagi et al. 1996).

DDT resistance in *Ae. albopictus* has also been reported in China (Neng et al. 1992), Central Africa (Kamgang et al. 2011), the United States of America (Marcombe et al. 2014), and Malaysia (Ishak et al. 2015). Widespread DDT resistance in *Ae. albopictus* was reported in urban districts in China, while rural populations remained susceptible (Neng et al. 1992). Marcombe et al. (2014) found that USA populations of *Ae. albopictus* were broadly susceptible to currently available larvicides and adulticides, except for significant resistance to DDT in two Florida populations and in a New Jersey population. The above worldwide spread of DDT resistance in *Ae. albopictus* may be attributable to the repeated introductions of the species into the American, European, and African continents from tropical, sub-tropical, and temperate Asian countries, including Japan.

6.2. Pyrethroid and DDT Resistance Mechanisms in *Aedes albopictus*

The synergistic interactions of DEM (diethylmaleate, an inhibitor of glutathione S-transferase (GST) activity, diethyl (2z-2-butenedioate)), DEF (tribuphos, an inhibitor of esterase activity, S,S,S-tributyl phosphorotrithionate), DMC (an inhibitor of DDT dehydrogenase activity, 4-Chloro-α-(chlorophenyl)-methyl-benzenemethanol), and PBO (piperonyl butoxide, an inhibitor of cytochrome P450 monooxygenase activity, 5-[2-(2-butoxyethoxy)ethoxymethyl]-6-propyl-1,3-benzodoxole) with DDT and permethrin in *Ae. albopictus* adults were examined using two Nagasaki colonies (Chuou and Heiwa) that showed high resistance to both pyrethroids and DDT. The Yonaguni colony, susceptible to both insecticides, and the Higashikurume colony, susceptible to pyrethroids and resistant to DDT, were used as references. For *Ae. albopictus* adults, synergistic activity with DDT was highest for DEM in both the Chuou and Heiwa colonies, while DMC and PBO were not effective. On the other hand, DEM, DMC, and PBO were synergistic with DDT in the Higashikurume colony. Comparisons could not be made among these synergists with permethrin because all tests showed 100% mortality, although an observed decrease in KT_{50} in the Chuou and Heiwa colonies when treated with the synergists may have been caused by the synergists. Larval susceptibility of the Chuou and Heiwa colonies was compared with the Higashikurume and Yonaguni colonies. In all colonies, PBO was more effective than DEF in reducing the LC_{50} of *d*-T_{80}-allethrin. The synergism of DEM was not as high as that of PBO in the Chuou colony and the Heiwa colony (Pujiyati et al. 2013).

Mutations in the VGSC (I1011, L1014, V1016, and F1534) were detected in the larvae and not a single mutation was found either in the 20 colonies collected from Nagasaki City or the eight colonies collected from the other places in Japan (Pujiyati et al. 2013).

Kawada et al. (2010) suggested the possibility of cross-resistance between pyrethroids and DDT in *Ae. albopictus* collected from Nagasaki City, because they showed resistance to pyrethroids. The organized and massive larvicidal treatment of graveyard containers with DDT formulations in the 1950s is thought to be one of the main causes for the development of pyrethroid resistance in this species, given that no other massive treatments with pyrethroids were conducted in this area.

Kasai et al. (2007) also suggested the same hypothesis concerning pyrethroid resistance in *Culex pipiens pallens* Coquillet in Japan. Possible mechanisms that might cause cross-resistance between pyrethroids and DDT include mutations in the VGSC (*kdr*) and enhanced cytochrome P450 monooxygenase activity. In the above study, however, no *kdr* mutations were found in the *Ae. albopictus* colonies. The contribution of PBO in enhancing DDT efficacy was not notable, indicating that neither the *kdr* mutation nor cytochrome P450 is a major cause of detoxification of DDT. Enhancement of esterase activity, which is thought to be blocked by DEF, and enhancement of GST activity, which is blocked by DEM, were both suggested in DDT-resistant colonies. The toxicological roles of both the above enzymes, however, have not yet been recognized as DDT-resistance factors, although a few studies have suggested correlations between these enzymes and DDT resistance (Sarkar et al. 2009, Lumjuan et al. 2011, Ranson et al. 2011). Enhancement of DDT dehydrochlorinase, which is blocked by DMC, might be one of the main resistance factors in the Higashi-Kurume colony, but the contribution of this enzyme was low in the Chuou and Heiwa colonies. The contribution of synergists in reducing pyrethroid resistance in adult *Ae. albopictus* was not clear, because all synergists caused 100% mortality in the adult susceptibility tests. The decrease in KT_{50} in the treatments using DEF, DEM, and PBO may explain the effectiveness of the synergists with pyrethroid. On the other hand, PBO was the most effective in reducing the LC_{50} of *d*-T_{80}-allethrin in the larval stage, indicating that cytochrome P450 might be one of the main resistance factors. Heng et al. (1992) demonstrated that DDT dehydrochlorinase was one of the main detoxification factors in DDT-resistant *Ae. albopictus* in China. Recently, biochemical assays revealed a significant up-regulation of GSTs in DDT-resistant populations collected in Florida, USA, in both larval and adult stages (Marcombe et al. 2014). Those authors concluded that the observed DDT resistance in populations from Florida may indicate multiple introductions of this species into the USA, possibly from tropical populations.

This study highlights the wide distribution of pyrethroid tolerance or resistance in *Ae. albopictus* in Nagasaki City and the widespread DDT resistance in this species in Japan. In the first half of the 1950s, DDT and BHC were sprayed around the peripheries of graveyards or into flower vases on gravestones in Nagasaki City. Following the ban on DDT use in the 1970s, organophosphates such as diazinon, temephos, fenthion, and fenitrothion were used instead of DDT, and since then, no pyrethroid insecticide has been used for mosquito control. Moreover, an organized system for mosquito control in Nagasaki City has been canceled, in accordance with the revision of the Infectious Disease Prevention Law in 2000. The continuing pyrethroid resistance in *Ae. albopictus* populations in Nagasaki City is therefore attributable to the massive and organized use of DDT in the 1950s as a mosquito larvae control measure (Kawada et al. 2010). In the end, however, we could not establish a clear relationship among resistance factors that might explain the cross-resistance between DDT and pyrethroids in pyrethroid-resistant *Ae. albopictus* in Japan. It is clear that the *kdr* mutation did not occur in *Ae. albopictus* during the earlier widespread DDT treatment. It can be hypothesized that multiple metabolic factors might possibly play roles in the detoxification of pyrethroids, and some of them might be common to both DDT and pyrethroid detoxification mechanisms. Further biochemical study is necessary for clarifying the above hypothesis.

Pyrethroid resistance in *Ae. albopictus* is not a serious worldwide problem at present (Jirakanjanakit et al. 2007, Vontas et al. 2012, Ishak et al. 2015, etc.), and pyrethroids provide one of the most promising countermeasures for controlling mosquitoes. Development of pyrethroid resistance, therefore, will be a major hindrance to mosquito control programs. At present, there are no suitable chemical substitutes for pyrethroids. A regular monitoring system for insecticide susceptibility, including a simple biochemical evaluation system that can elucidate the modes of resistance, should be given priority in mosquito resistance management.

Conclusion

DDT resistance seems to have spread and become stable worldwide in mosquito populations despite the long interval since the ban of the insecticide. This may be true in the other insect populations as well. Genetic insecticide resistance factors that are beneficial to insects under pressure from xenobiotic insecticides sometimes impose negative costs on those insects, and such genetic factors are diminished in the course of time (Coustau et al. 2000, Hall et al. 2004). Platt et al. (2015) demonstrated that heterozygous *kdr* (L1014F) males from multiple resistant populations of *Anopheles coluzzii* (*An. gambiae* s.l.) were more likely to mate than homozygous resistant males, resulting in a negative impact on fitness from the *kdr* gene. In contrast, no difference in fitness was observed for the main metabolic resistance genes (Ace-1 mutation). Similarly, Brito et al. (2013) showed that *kdr* mutations (F1534C and V1016I) in *Ae. aegypti* have a high fitness cost in larval development, egg production, and fertility.

In DDT-resistant *Drosophila melanogaster* (DDT-R), the DDT resistance is approaching fixation globally, long after the withdrawal of DDT (Catania et al. 2004). Surprising results were reported by McCart et al. (2005), who found that the above DDT-R locus actually increases both adult fecundity and the viability of eggs and larvae, as well as speeding both larval and pupal development. The authors demonstrated that the single DDT-R allele, or a closely linked modifier, conferring a female-linked advantage in the absence of insecticide selection, together with the continued use of other insecticides to which DDT-R confers cross-resistance, may help explain the fact that DDT-R is approaching global fixation. The reason the DDT resistance gene has no fitness cost in DDT-resistant mosquitos or other insect populations is unknown. We can easily imagine, however, that pyrethroid resistance in mosquito populations is now going to spread worldwide and become fixed in the same manner as DDT resistance. Pyrethroid resistance in vector mosquitoes is now the greatest obstacle for the success of malaria and other mosquito-borne infectious disease control programs. Acceleration of molecular-based studies that identify the insecticide resistance mechanisms and break up the development of insecticide resistance are urgently needed, as are developmental studies of new insecticides with novel modes of action and new designs for controlling devices that allow no countermeasures by vector mosquitoes.

ACKNOWLEDGMENTS

These studies were supported by the funding from Center for Infectious Disease Research in Asia and Africa: Nagasaki University, Japan and Program of Japan Initiative for Global Research Network on Infectious Diseases (J-GRID), MEXT, Japan.

REFERENCES

Agadzi, V. K., Boatin, B. A., Appawu, M. A., Mingle, A. A. & Addy, P. A. (1984). Yellow fever in Ghana, 1977-80. *Bull WHO*, *62*, 577–583.

Alvarez, L. C., Ponce, G., Saavedra-Rodriguez, K., Lopez, B. & Flores, A. E. (2015). Frequency of V1016I and F1534C mutations in the voltage-gated sodium channel gene in *Aedes aegypti* in Venezuela. *Pest Manag Sci.*, *71*, 863–869. doi: 10.1002/ps.3846.

Appawu, M., Dadzie, S., Abdul, H., Asmah, H., Boakye, D., Wilson, M. & Ofori-Adjei, D. (2006). Surveillance of viral haemorrhagic fevers in Ghana: Entomological assessment of the risk of transmission in the northern regions. *Ghana Med J*, *40*, 137-141. http:// www.ncbi.nlm. nih.gov/pmc/articles/PMC1868007/.

Aung, T. T., Win, S. & Aung, S. (1996). Status report on epidemiology of dengue/dengue haemorrhagic fever in Myanmar, 1995. *Dengue Bull*, *20*, 41–45.

Binka, F. N., Kubaje, A., Adjuik, M., William, s. L. A., Lengeler, C., Maude, G. H., Arma, G. E., Kajihra, B., Adiama, J. H., Lengeler, C., Maude, G. H., Armah, G. E., Kajihara, B., Adiamah, J. H. & Smith, P. G. (1996). Impact of permethrin treated bed nets in child mortality in Kassena-Nankana district, Ghana: a randomized controlled trial. *Trop Med Int Health*, *1*, 147–154. doi: http://dx.doi.org/10.1111/j.1365-3156. 1996. tb00020.x.

Bonilauri, P., Bellini, R., Calzolari, M., Angelini, R., Venturi, L., Fallacara, F., Cordioli, P., Angelini, P., Venturelli, C., Merialdi, G. & Dottori, M. (2008). Chikungunya virus in *Aedes albopictus*, Italy. *Emerg Infect Dis*, *14*, 852–854. http://wwwnc.cdc.gov/eid/article/14/5/07-1144_article.

Brengues, C., Hawkes, N. J., Chandre, F., McCaroll, L., Duchon, S., Guillet, P., Manguin, S., Morgan, J. C. & Hemingway, J. (2003). Pyrethroid and DDT cross-resistance in *Aedes aegypti* is correlated with novel mutations in the voltage-gated sodium channel gene. *Med Vet Entomol*, *17*, 87–94. doi: 10.1046/j.1365-2915.2003.00412.x.

Brito, L. P., Linss, J. G., Lima-Camara, T. N., Belinato, T. A., Peixoto, A. A., Lima, J. B., Valle, D. & Martins, A. J. (2013). Assessing the effects of *Aedes aegypti kdr* mutations on pyrethroid resistance and its fitness cost. *PLoS One*, *8*, e60878. doi: 10.1371/journal.pone.0060878.

Bruce-Chwatt, L. J. (1985). *Essential Malariology*. New York, John Wiley & Sons.

Catania, F., Kauer, M. O., Daborn, P. J., Yen, J. L., ffrench-Constant, R. H. & Schlotterer, C. (2004). World-wide survey of an *Accord* insertion and its association with DDT resistance in *Drosophila melanogaster*. *Mol Ecol*, *13*, 2491–2504. doi: 10.1111/j.1365-294X.2004.02263.x.

Coustau, C., Chevillon, C. & ffrench-Constant, R. H. (2000). Resistance to xenobiotics and parasites: can we count the cost? *Trends Ecol Evol*, *15*, 378–383. doi:10.1016/S0169-5347(00)01929-7.

Du, Y., Nomura, Y., Satara, G., Hua, Z., Nauen, R., He, S. Y., Zhorov, B. S. & Dong, K. (2013). Molecular evidence for dual pyrethroid-receptor sites on a mosquito sodium channel. *PNAS*, *110*, 11785–11790. doi: 10.1073/pnas.1305118110.

Edman, J. D., Kittayapong, P., Linthicum, K. & Scott, T. (1997). Attractant resting boxes for rapid collection and surveillance of *Aedes aegypti* (L.) inside houses. *J Am Mosq Control Assoc*, *13*, 24–27.

Enayati, A. A., Vatandoost, H., Ladonni, H., Townson, H. & Hemingway, J. (2003). Molecular evidence for a *kdr*-like pyrethroid resistance mechanism in the malaria vector mosquito *Anopheles stephensi*. *J Med Vet Entomol*, *17*, 138–144. doi: 10.1046/j.1365-2915.2003.00418.x.

Erhart, A., Thang, N. D., Hung, N. Q., Toi, L. V., Hung, L. X., Tuy, T. Q., Cong, L. D., Speybroecck, N., Coosemans, M. & D'Alessandro, U. (2004). Forest malaria in Vietnam: A challenge for control. *Am J Trop Med Hyg*, *70*, 110–118. doi: http://www.ajtmh.org/content/70/2/110.full.pdf+html.

Erhart, A., Thang, N. D., Ky, P. V., Tinh, T. T., Overmeir, C. V., Speybroeck, N., Obsomer, V., Hung, L. X., Thuan, L. K., Coosemans, M. & D'alessandro, U. (2005). Epidemiology of forest malaria in central Vietnam: a large scale cross-sectional survey. *Malaria Journal*, *4*, 58. doi: 10.1186/1475-2875-4-58.

Failloux, A. B., Vazeille, M. & Rodhain, F. (2002). Geographic genetic variation in populations of the dengue virus vector *Aedes aegypti*. *J Mol Evol*, *55*, 653–663. doi: 10.1007/s00239-002-2360-y.

Fanello, C., Petrarca, V., Della Torre, A., Santolamazza, F., Dolo, G., Coulibaly, M., Alloueche, M., Curtis, C. F., Touré, Y. T. & Coluzzi, M. (2003). The pyrethroid knock-down resistance gene in the *Anopheles gambiae* complex in Mali and further indication of incipient speciation within *An. gambiae* s.s. *Insect Mol Biol*, *12*, 241–245. doi: 10.1046/j.1365-2583.2003.00407.x.

Diabaté, A., Baldet, T., Chandre, F., Guiguemdé, R. T., Brengues, C., Guillet, P., Hemingway, J. & Hougard, J. M. (2002). First report of the *kdr* mutation in *Anopheles gambiae* M form from Burkina Faso, west Africa. *Parassitologia*, *44*, 157–158.

Fianko, J. R., Donkor, A., Lowor, S. T. & Yeboah, P. O. (2011). Agrochemicals and the Ghanaian environment, a review. *J Environ Protect*, *2*, 221–230. doi:10.4236/jep.2011.23026.

Gentry, J. W. & Hubert, A. A. (1957). Resistance of Culex quinquefasciatus to chlorinated hydrocarbons on Okinawa. *Mosq News*, *17*, 92–93.

Gubler, D. J. (1997). Dengue and dengue hemorrhagic fever: Dengue and dengue hemorrhagic fever:its history and resurgence as a global public health problem. (Gubler DJ, Kuno G, editors). Oxon and New York, Cab International, pp. 1–22.

Guillet, P., Alnwick, D., Cham, M. K., Neira, M., Zaim, M., Heymann, D. & Mukelabai, K. (2001). Long-lasting treated mosquito nets: a breakthrough in malaria prevention. *Bull WHO*, *79*, 998. doi: http:// dx.doi.org/ 10.1590/S0042-96862001001000017.

Haibara, E. (2012). Challengers for malaria eradication. Yaima Bunko 13, Tokyo, Nanzan-sha Press (in Japanese).

Hall, R. J., Gubbins, S. & Gilligan, C. A. (2004). Invasion of drug and pesticide resistance is determined by a trade-off between treatment efficacy and relative fitness. *Bull Math Biol*, *66*, 825–840. doi: 10.1016/ j.bulm.2003.11.006.

Harris, A. F., Rajatileka, S. & Ranson, H. (2010). Pyrethroid resistance in *Aedes aegypti* from Grand Cayman. *Am J Trop Med Hyg*, *83*, 277–284. doi:10.4269/ajtmh.2010.09-0623.

Hawley, W. A. (1988). The biology of *Aedes albopictus*. *J Am Mosq Control Assoc*, (Suppl. 1), *4*, 1–39.

Higa, Y., Tsuda, Y., Tuno, N. & Takagi, M. (2001). Preliminary field experiments on exophagy of *Aedes albopictus* (Diptera: Culicidae) in peridomestic habitat. *Med Entomol Zool*, *52*, 105–116.

Higa, Y., Nguyen, Y. T., Kawada, H., Tran, S. H., Nguyen, H. T. & Takagi, M. (2010). Geographic distribution of *Aedes aegypti* and *Aedes albopictus* collected from used tires in Vietnam. *J Am Mosq Control Assoc*, *26*, 1–9. doi: 10.2987/09-5945.1.

Hirata, K., Komagata, O., Itokawa, K., Yamamoto, A., Tomita, T. & Kasai, S. (2014). A single crossing-over event in voltage-sensitive Na^+ channel genes may cause critical failure of dengue mosquito control by insecticides. *PLoS Negl Trop Dis*, *8*, e3085. doi: 10.1371/ journal. pntd. 0003085.

Hosaka, H. (1992). A case study of U.S. Occupation in Okinawa - Focusing on the preventive medicine and the public health activities-. *Bulletin of the College of Law and Letters, University of the Ryukyus. Sociology (34)*, 1–33 (in Japanese). http://ir.lib.u-ryukyu.ac.jp/handle/123456789/12927.

Hotta, S. (1998). Dengue vector mosquitoes in Japan. The role of *Aedes albopictus* and *Aedes aegypti* in the 1942-1944. *Med Entomol Zool*, *43*, 276–274.

Huber, K., Luu, L. L., Tran, H. H., Tran, K. T. Rodhain, F. & Failloux, A. B. (2003). *Aedes aegypti* in south Vietnam: Ecology, genetic structure, vectorial competence and resistance to insecticides. *South East Asian J Trop Med Public Health*, *34*, 81–86.

Hung, L. Q., de Vries, P. J., Giao, P. T., Nam, V. S., Binh, T. Q., Chong, M. T., Quoc, N. T. T. A., Thanh, T. N., Hung, L. N. & Kager, P. A. (2002). Control of malaria: a successful experience from Viet Nam. *Bull WHO*, *80*, 660–666. doi: http:// www.ncbi.nlm.nih.gov/ pmc/articles/ PMC2567582/.

Ishak, H., Miyagi, I., Toma, T. & Kamimura, K. (1997). Breeding habitats of *Aedes aegypti* (L.) and *Aedes albopictus* (Skuse) in villages of Barru, South Sulawesi, Indonesia. *Southeast Asian J Trop Med Public Health*, *28*, 844–850. doi: 10.1186/s13071-015-0797-2.

Ishak, I. H., Jaal, Z., Ranson, H. & Wondji, C. S. (2015). Contrasting patterns of insecticide resistance and knockdown resistance (*kdr*) in the dengue vectors *Aedes aegypti* and *Aedes albopictus* from Malaysia. *Parasit Vect*, *8*, 181. doi: 10.1186/s13071-015-0797-2.

Jirakanjanakit, N., Rongnoparut, P., Saengtharatip, S., Chareonviriyaphap, T., Duchon, S., Bellec, C. & Yoksani, S. (2007). Insecticide Susceptible/Resistance Status in *Aedes* (Stegomyia) *aegypti* and *Aedes* (Stegomyia) *albopictus* (Diptera: Culicidae) in Thailand During 2003-2005. *J Econ Entomol*, *100*, 545–550. doi: http:// dx.doi.org/ 10.1093/ jee/100.2.545.

Kamgang, B., Marcombe, S., Chandre, F., Nchoutpouen, E., Nwane, P., Etang, J., Corbel, V. & Paupy, C. (2011). Insecticide susceptibility of *Aedes aegypti* and *Aedes albopictus* in Central Africa. *Parasit Vect*, *4*, 79. http:// www. parasitesandvectors.com/content/4/1/79.

Kasai, S., Shono, T., Komagata, O., Tsuda, Y., Kobayashi, M., Motoki, M., Kashima, I., Tanikawa, T., Yoshida, M., Tanaka, I., Shinjo, G., Hashimoto, T., Ishikawa, T., Takahashi, T., Higa, Y. & Tomita, T. (2007). Insecticide resistance in potential vector mosquitoes for West Nile virus in Japan. *J Med Entomol*, *44*, 822–829. doi: http:// dx.doi.org/ 10.1093/ jmedent/ 44.5.822.

Kasai, S., Ng, L. C., Lam-Phua, S. G., Tang, C. S., Itokawa, K., Komagata, O., Kobayashi, M. & Tomita, T. (2011). First detection of a putative knockdown resistance gene in major mosquito vector, *Aedes albopictus*. *Jpn J Infect Dis*, *64*, 217–221. http://www0.nih.go.jp/JJID/64/217.pdf.

Kawada, H., Higa, Y., Nguyen, Y. T., Tran, S. H., Nguyen, H. T. & Takagi, M. (2009a). Nationwide investigation of the pyrethroid susceptibility of mosquito larvae collected from used tires in Vietnam. *PLoS Negl Trop Dis*, *3*, e391. doi: 10.1371/journal.pntd.0000391.

Kawada, H., Higa, Y., Komagata, O., Kasai, S., Tomita, T., Yen, N. T., Loan, L. L., Sánchez, R. A. & Takagi, M. (2009b). Widespread distribution of a newly found point mutation in voltage-gated sodium channel in pyrethroid-resistant *Aedes aegypti* populations in Vietnam. *PLoS Negl Trop Dis*, *3*, e527. doi: 10.1371/journal.pntd.0000527.

Kawada, H., Maekawa, Y., Abe, M., Ohashi, K., Ohba, S. & Takagi, M. (2010). Spatial distribution and pyrethroid susceptibility of mosquito larvae collected from catch basins in parks in Nagasaki city, Nagasaki, Japan. *Jpn J Infect Dis*, *63*, 19–24. http://naosite.lb.nagasaki-u.ac.jp/ dspace/ bitstream/10069/23060/1/JJID63_19.pdf.

Kawada, H., Oo, S. Z., Thaung, S., Kawashima, E., Maung, Y. N., Thu, H. M., Thant, K. Z. & Minakawa, N. (2014). Co-occurrence of point mutations in the voltage-gated sodium channel of pyrethroid-resistant *Aedes aegypti* populations in Myanmar. *PLoS Negl Trop Dis*, *8*, e3032. doi: 10.1371/journal.pntd.0003032.

Kawada, H., Higa, Y., Futami, K., Muranami, Y., Kawashima, E., Osei, J. H. N., Dadzie, S., Sakyi, K. Y., de Souza, D., Appawu, M., Ohta, N., Suzuki, T. & Minakawa, N. (2016). Discovery of point mutations in the voltage-gated sodium channel from African *Aedes aegypti* populations: Potential phylogenetic reasons for gene introgression. *PLoS Negl Trop Dis*, *10*, e0004780.

Kliot, A. & Ghanim, M. (2012). Fitness costs associated with insecticide resistance. *Pest Manag Sci*, *68*, 1431–1437. doi: 10.1002/ps.3395.

Kristan, M., Fleischmann, H., della Torre, A., Stich, A. & Curtis, C. F. (2003). Pyrethroid resistance/susceptibility and differential urban/rural distribution of *Anopheles arabiensis* and *An. gambiae* s. s. malaria vectors in Nigeria and Ghana. *Med Vet Entomol*, *17*, 326–332. doi: 10.1046/j.1365-2915.2003.00449.x.

Kudom, A. A., Mensah, B. A. & Agyemang, T. K. (2012). Characterization of mosquito larval habitats and assessment of insecticide-resistance status of *Anopheles gambiae* senso lato in urban areas in southwestern Ghana. *J Vector Ecol*, *37*, 77–82. doi: 10.1111/j.1948-7134.2012.00202.x.

Kurihara, T. (2002). Malaria vectors in Japan: except Nansei Islands. *Med Entomol Zool*, *53 (Supplement 2)*, 1–28 (in Japanese).

Kushwah, R. B., Dykes, C. L., Kapoor, N., Adak, T. & Singh, O. P. (2015). Pyrethroid-resistance and presence of two knockdown resistance (*kdr*) mutations, F1534C and a novel mutation T1520I, in Indian *Aedes aegypti*. *PLoS Negl Trop Dis*, *9*, e3332. doi: 10.1371/journal.pntd.0003332.

Kutsuna, S., Kato, Y., Moi, M. L., Kotaki, A., Ota, M., Shinohara, K., Kobayashi, T., Yamamoto, K., Fujiya, Y., Mawatari, M., Sato, T., Kunimatsu, J., Takeshita, N., Hayakawa, K., Kanagawa, S., Takasaki, T. & Ohmagari, N. (2015). Autochthonous dengue fever, Tokyo, Japan, 2014. *Emerg Infect Dis*, *21*, 517–520. doi: 10.3201/eid2103.141662.

Lengeler, C., Cattani, J. & De Savigny, D. (1996). Net gain. A new method for preventing malaria deaths (International Development Research Center (Canada), WHO, pp. 189.

Linss, J. G., Brito, L. P., Garcia, G. A., Araki, A. S., Bruno, R. V., Lima, J. B., Valle, D. & Martins, A. J. (2014). Distribution and dissemination of the Val1016Ile and Phe1534Cys *Kdr* mutations in *Aedes aegypti* Brazilian natural populations. *Parasit Vect*, *7*, 25. doi: 10.1186/1756-3305-7-25.

Lumjuan, N., Rajatileka, S., Changsom, D., Wicheer, J., Leelapat, P., Prapanthadara, L. A., Somboon, P., Lycett, G. & Ranso, n. H. (2011). The role of the *Aedes aegypti* Epsilon glutathione transferases in conferring resistance to DDT and pyrethroid insecticides. *Insect Biochem Mol Biol*, *41*, 203–209. doi:10.1016/j.ibmb.2010.12.005.

Marcombe, S., Mathieu, R. B., Pocquet, N., Riaz, M. A., Poupardin, R., Sélior, S., Darriet, F., Reynaud, S., Yébakima, A., Corbel, V., David, J. P. & Chandre, F. (2012). Insecticide resistance in the dengue vector *Aedes aegypti* from Martinique: Distribution, mechanisms and relations with environmental factors. *PLoS One*, *7*, e30989. doi: 10.1371/ journal.pone. 0030989.

Marcombe, S., Farajollahi, A., Healy, S. P., Clark, G. G. & Fonseca, D. M. (2014). Insecticide resistance status of United States populations of *Aedes albopictus* and mechanisms involved. *PLoS One*, *9*, e101992. doi:10. 1371/journal.pone.0101992.

Martins, A. J., Lima, J. B. P., Peixoto, A. A. & Valle, D. (2009). Frequency of Val1016Ile mutation in the voltage-gated sodium channel gene of *Aedes aegypti* Brazilian populations. *Trop Med Int Health*, *14*, 1351–1355. doi: 10.1111/j.1365-3156.2009.02378.x.

Martinez-Torres, D., Chandre, F., Williamson, M. S., Darret, F., Bergé, J. B., Devonshire, A. L., Guillet, P., Pasteur, N. & Pauron, D. (1998). Molecular characterization of pyrethroid knockdown resistance (*kdr*) in the major malaria vector *Anopheles gambiae* s.s. *Insect Mol Biol*, *7*, 179–184. doi: 10.1046/j.1365-2583.1998.72062.x.

Martinez-Torres, D., Chevillon, C., Brun-Barale, A., Bergé, J. B., Pasteur, N. & Pauron, D. (1999). Voltage-dependent Na^+ channels in pyrethroid-resistant *Culex pipiens* L. mosquitoes. *Pestic Sci*, *55*, 1012–1020. doi: 10.1002/(SICI)1096-9063(199910)55:10<1012::AID-PS39>3.0.CO; 2-5/.

Mattingly, P. F. (1957). Genetical aspects of the *Aedes aegypti* problem, I. Taxonomy and bionomics. *Ann Trop Med Parasitol*, *51*, 392–408.

McCart, C., Buckling, A. & ffrench-Constant, R. H. (2005). DDT resistance in flies carries no cost. *Current Biol*, *15*, R587–R589. doi:10.1016/ j.cub. 2005.07.054.

Miyagi, I., Toma, T., Malenganisho, W. L. & Uza, M. (1996). Historical review of mosquito control as a component of malaria eradication program in the Ryukyu Archipelago. *Southeast Asian J Trop Med Public Health*, *27*, 498–511.

Miyagi, I., Toma, T., Zyasu, N. & Takashita, Y. (1989). Insecticide susceptibility of *Culex quinquefasciatus* larvae (Diptera: Culicidae) in Okinawa Prefecture, Japan in 1989. *Jpn J Sanit Zool*, *45*, 7–11.

Morishita, K. (1963). *Parasitological Research in Japan III*, Tokyo, Meguro Kiseityu Kan (in Japanese).

Nam, N. V., de Vries, P. J., Toi, L. V. & Nagelkerke, N. (2005). Malaria control in Vietnam: the Binh Thuan experience. *Trop Med Int Health*, *10*, 357–365. doi: 10.1111/j.1365-3156.2005.01387.x.

Neng, W., Yan, X., Fuming, H. & Dazong, C. (1992). Susceptibility of *Aedes albopictus* from China to insecticides, and mechanism of DDT resistance. *J Am Mosq Control Assoc*, *8*, 394–397. http:// www.biodiversitylibrary.org/content/part/JAMCA/JAMCA_V08_N4_P394-397.pdf.

Nevill, C. G., Some, E. S., Mung'ala, V. O., Mutemi, W., New, L., Marsh, K., Lengeler, C. & Snow, R. W. (1996). Insecticide-treated bed nets reduce mortality and severe morbidity from among children on the Kenyan Coast. *Trop Med Int Health*, *1*, 139–146. doi: http://dx.doi.org/10.1111/j.1365-3156.1996.tb00019.x.

Nguyen, T. H. (1994). Country report on dengue fever/dengue haemorrhagic fever and Japanese encephalitis in Vietnam. *Trop Med*, *36*, 170–176.

Ohtsuru, M. (1958). Why the postwar imported malaria did not cause indigenous infection? In T. Ogata (Eds.), *Chihou-byou kenkyu no doukou* No. 22, Tokyo, Kinbara Press, pp 107–137 (in Japnese).

Oo, T. T., Storch, V., Madon, M. B. & Becker, N. (2011). Factors influencing the seasonal abundance of *Aedes* (Stegomyia) *aegypti* and the control strategy of dengue and dengue hemorrhagic fever in Thanlyin Township, Yangon City, Myanmar. *Trop Biomed*, *28*, 302–311. http:// www.msptm.org/files/302_-_311_Oo_T_T.pdf.

Pang, S. C., Chiang, L. P., Tan, C. H., Vythilingam, I., Lam-Phua, S. G. & Ng, L. C. (2015). Low efficacy of deltamethrin-treated net against Singapore *Aedes aegypti* is associated with *kdr*-type resistance. *Trop Biomed*, *32*, 140–150.

Pethuan, S., Jirakanjanakit, N., Saengtharatip, S., Chareonviriyaphap, T., Kaewpa, D. & Rongnoparut, P. (2007). Biochemical studies of insecticide resistance in *Aedes* (Stegomyia) *aegypti* and *Aedes* (Stegomyia) *albopictus* (Diptera: Culicidae) in Thailand. *Trop Biomed, 24*, 7–15. doi: http://www.msptm.org/files/07_-_15_Pethuan_S.pdf.

Ping, L. T., Yatiman, R. & Gek, L. S. (2001). Susceptibility of adult field strains of *Aedes aegypti* and *Aedes albopictus* in Singapore to pirimiphos-methyl and permethrin. *J Am Mosq Control Assoc, 17*, 144–146.

Platt, N., Kwiatkowska, R. M., Irving, H., Diabaté, A., Dabire, R. & Wondji, C. S. (2015). Target-site resistance mutations (*kdr* and RDL), but not metabolic resistance, negatively impact male mating competiveness in the malaria vector *Anopheles gambiae*. *Heredity, 115*, 243–52. doi: 10.1038/hdy.2015.33.

Ponlawat, A., Scott, J. G. & Harrington, L. C. (2005). Insecticide Susceptibility of *Aedes aegypti* and *Aedes albopictus* across Thailand. *J Med Entomol, 42*, 821–825. doi: http:// dx.doi.org/ 10.1093/ jmedent/ 42.5.821.

Powers, A. M. & Logue, C. H. (2007). Changing patterns of chikungunya virus: re-emergence of a zoonotic arbovirus. *J Gen Virol, 88*, 2363–2377. doi: 10.1099/vir.0.82858-0.

Pujiyati, E., Kawada, H., Sunahara, T., Kasai, S. & Minakawa, N. (2013). Pyrethroid resistance status of *Aedes albopictus* (Skuse) collected in Nagasaki City, Japan. *Jpn J Environ Entomol Zool, 24*, 143–153.

Rajatileka, S., Black IV, W. C., Saavedra-Rodriguez, K., Trongtokit, Y., Apiwathnasorn, C., McCall, P. J. & Ranson, H, (2008). Development and application of a simple colorimetric assay reveals widespread distribution of sodium channel mutations in Thai population of *Aedes aegypti*. *Acta Trop, 108*, 54–57. doi: 10.1016/j.actatropica.2008.08.004.

Ranson, H., Rossiter, L., Ortelli, F., Jensen, B., Wang, X., Roth, C. W., Collins, F. H. & Hemingway, J. (2001). Identification of a novel class of insect glutathione S-transferases involved in resistance to DDT in the malaria vector *Anopheles gambiae*. *Biochem J, 359*, 295–304. doi: 10.1042/bj3590295.

Reiter, P. (1998). *Aedes albopictus* and the world trade in used tires, 1988-1995: The shape of things to come? *J Am Mosq Control Assoc, 14*, 83–94. http://www.biodiversitylibrary.org/content/part/JAMCA/JAMCA_V14_N1_P083-094.pdf

Reiter, P. & Sprenger, D. (1987). The used tire trade: A mechanism for the worldwide dispersal of container breeding mosquitoes. *J Am Mosq Control Assoc*, *3*, 494–501.

Riveron, J. M., Yunta, C., Ibrahim, S. S., Djouaka, R., Irving, H., Menze, B. D., Ismail, H. M., Hemingway, J., Ranson, H., Albert, A. & Wondji, C. S. (2014). A single mutation in the GSTe2 gene allows tracking of metabolically based insecticide resistance in a major malaria vector. *Genome Biol*, *15*, R27. http://genomebiology.com/2014/15/2/R27.

Rodhain, F. & Rosen, L. (1997). Dengue and dengue hemorrhagic fever: Mosquito vector and dengue virus-vector relationships. (Gubler DJ, Kuno G, editors). Oxon and New York, Cab International, pp. 45–60.

Saavedra-Rodriguez, K., Urdaneta-Marquez, L., Rajatileka, S., Moulton, M., Flores, A. E., Fernandez-Salas, I., Bisset, J., Rodriguez, M., Mccall, P. J., Donnelly, M. J., Ranson, H., Hemingway, J. & Black IV, W. C. (2007). A mutation in the voltage-gated sodium channel gene associated with pyrethroid resistance in Latin American *Aedes aegypti*. *Insect Mol Biol*, *16*, 785–798. doi: 10.1111/j.1365-2583.2007.00774.x.

Sarkar, M., Bhattacharya, I. K., Borkotoki, A., Goswami, D., Rhabha, B., Baruah, I. & Srivastava, R. B. (2009). Insecticide resistance and detoxifying enzyme activity in the principal bancroftian filariasis vector, *Culex quenquefasciatus*, in northeastern India. *Med Ve. Entomol*, *23*, 122–131. doi: 10.1111/j.1365-2915.2009.00805.x.

Sawada, T. (1949). Postwar malaria. *Nippon Naikagakkai Zassi*, *38*, 1–14 (in Japanese).

Seixas, G., Salgueiro, P., Silva, A. C., Campos, M., Spenassatto, C., Reyes-Lugo, M., Novo, M. T., Ribolla, P. E., Silva Pinto, J. P. & Sousa, C. A. (2013). *Aedes aegypti* on Madeira Island (Portugal): genetic variation of a recently introduced dengue vector. *Mem Inst Oswaldo Cruz*, *108*, Suppl 1, 3–10. doi: 10.1590/0074-0276130386.

Siller, Q., Ponce, G., Lozano, S. & Flores, A. E. (2011). Update on the frequency of Ile1016 mutation in voltage-gated sodium channel gene of *Aedes aegypti* in Mexico. *J Am Mosq Control Assoc*, *27*, 357–362. doi: http://dx.doi.org/10.2987/11-6149.1.

Srisawat, R., Komalamisra, N., Eshita, Y., Zheng, M., Ono, K., Itoh, T. Q., Matsumoto, A., Petmitr, S. & Rongsriyam, Y. (2010). Point mutations in domain II of the voltage-gated sodium channel gene in deltamethrin-resistant *Aedes aegypti* (Diptera: Culicidae). *Appl Entomol Zool*, *45*, 275–282. http://doi.org/10.1303/aez.2010.275.

Stanton, A. T. (1920). The mosquitoes of far East ports with special reference to the prevalence of *Stegomyia fasciata*. *Bull Entomol Res*, *10*, 333–334.

Stenhouse, S. A., Plernsub, S., Yanola, J., Lumjuan, N., Dantrakool, A., Choochote, W. & Somboon, P. (2013). Detection of the V1016G mutation in the voltage-gated sodium channel gene of *Aedes aegypti* (Diptera: Culicidae) by allele-specific PCR assay, and its distribution and effect on deltamethrin resistance in Thailand. *Parasit Vect*, *6*, 253. doi: 10.1186/1756-3305-6-253.

Stoler, J., Delimini, R. K., Bonney, J. H. K., Oduro, A. R., Owusu-Agyei, S., Fobil, J. N. & Awandare, G. A. (2015). Evidence of recent dengue exposure among malaria parasite-positive children in three urban centers in Ghana. *Am J Trop Med Hyg*, *92*, 497–500. doi: 10.4269/ajtmh.14-0678.

Surtees, G. (1967). The distribution, density and seasonal prevalence of *Aedes aegypti* in West Africa. *Bull WHO*, *36*, 539–540.

Suzuki, T. (1962). Susceptibility or resistance to insecticides in lesser house fly, Sarcophaged fly, and two species of blowfly in Japan. *Jpn J Exp Med*, *32*, 309–313.

Suzuki, T. (1963). Insecticide resistance in flies, mosquitoes and cockroaches in Japan evaluated by topical application tests, with special reference to the susceptibility levels of the insects. *Jpn J Exp Med*, *33*, 69–83.

Suzuki, T. & Mizutani, K. (1962). Studies on insecticide resistance in mosquitoes of Japan. *Jpn J Exp Med*, *32*, 297–308.

Takasaki, T. (2011). Imported dengue fever/dengue hemorrhagic fever cases in Japan. *Trop Med Health*, *39*, S13–5. doi: 10.2149/tmh.2011-S07.

Thang, N. D., Erhart, A., Speybroeck, N., Hung, L. X., Thuan, L. K., Hung, C. T., Ky, P. V., Coosemans, M. & D'Alessandro, U. (2008). Malaria in central Vietnam: analysis of risk factors by multivariate analysis and classification tree models. *Malar J*, *7*, 28. doi: 10.1186/1475-2875-7-28.

Thaung, U., Ming, C. K. & Thein, M. (1975). Insecticide susceptibility of some vector fleas and mosquitoes in Burma. *Southeast Asian J Trop Med Public Health*, *6*, 555–61.

Toma, T., Miyagi, I., Chinen, T. & Hatazoe, H. (1992). Insecticidal Susceptibility of *Aedes albopictus* larvae in Different Island of Okinawa Perfecture, Japan. *Jpn J Sanit Zool*, *43*, 331–336.

Toma, T., Miyagi, I., Kishimoto, T., Higa Y., Hatazoe, H. & Zayasu, N. (1993). Insecticidal resistance of *Culex tritaeniorhynchus* Giles (Diptera : Culicidae) in Okinawa Prefecture, Japan, in 1970-1972 and 1989. *Jpn J Sanit Zool*, *44*, 263–269.

Toma, T., Miyagi, I., Malenganisho, W. L. M. & Tamashiro, M. (1998). Insecticide susceptibility of *An. minimus*, *An. saperoi* and *An. sinensis* larvae (Diptera: Culicidae) in Okinawa prefecture, Japan, in 1993 and 1994. *Med Entomol ZooL*, *49*, 227–230.

Tsuda, Y., Suwonkerd, W., Chawprom, S., Prajakwong, S. & Takagi, M. (2006). Different spatial distribution of *Aedes aegypti* and *Aedes albopictus* along an urban-rural gradient and the relating environmental factors examined in three villages in northern Thailand. *J Am Mosq Control Assoc*, *22*, 222–228.

Tsukamoto, M., Ogaki, M. & Kobayashi, H. (1957). Malaria control and the development of DDT resistant insects in Hikone City, Japan. *Jpn J Sanit Zool*, *8*, 118–122.

Umeda, K., Shono, T., Hirano, M. & Takahashi, M. (1990). Reduced nerve sensitivity as a resistant mechanism to a pyrethroid in *Culex tritaeniorhynchus* from Okinawa. *J Pestic Sci*, *15*, 599–601.

van den Berg, H., Zaim, M., Yadav, R. S., Soares, A., Ameneshewa, B., Mnzava, A., Hii, J., Dash, A. P. & Ejov, M. (2012). Global trends in the use of insecticides for vector-borne disease control. *Environ Health Perspect*, *120*, 577–582. doi: http://dx.doi.org/10.1289/ehp.1104340.

Vera-Maloof, F. Z., Saavedra-Rodriguez, K., Elizondo-Quiroga, A. E., Lozano-Fuentes, S. & Black IV, W. C. (2015). Coevolution of the Ile1,016 and Cys1,534 mutations in the voltage gated sodium channel gene of *Aedes aegypti* in Mexico. *PLoS Negl Trop Dis*, *9*, e0004263. http://dx.doi.org/10.1371/journal.pntd.0004263.

Verlé, P., Lieu, T. T. T., Kongs, A., der Stuyft, P. V. & Coosemans, M. (1999). Control of malaria vectors: cost analysis in a province of northern Vietnam. *Trop Med Int Health*, *4*, 139–145.

Vontas, J., Kioulos, E., Pavlidi, N., Morou, N., della Torre, A. & Ranson, H. (2012). Insecticide resistance in the major dengue vectors *Aedes albopictus* and *Aedes aegypti*. *Pestic Biochem Physiol*, *104*, 126–131. doi:10.1016/ j.pestbp.2012.05.008.

Vu, D. H., Nguyen, T. B. N., Do, T. H. & Nguyen, T. B. L. (2004). Susceptibility of *Aedes aegypti* to insecticides in Viet Nam. *Dengue Bull*, *28*, 179–183.

WHO, (1986). Prevention and control of yellow fever in Africa. Prevention and control handbooks. http://www.who.int/iris/handle/10665/39154.

Yanola, J., Somboon, P. & Prapanthadara, L. (2008). A novel point mutation in the *Aedes aegypti* voltage-gated sodium channel gene associated with permethrin resistance. *The 2nd International Conference on Dengue and Dengue Haemorhagic Fever*, Oct. 15–17, 2008, Phuket, Thailand.

Yanola, J., Somboon, P., Walton, C., Nachaiwieng, W., Somwang, P. & Prapanthadara, L. (2011). High-throughput assays for detection of the F1534C mutation in the voltage-gated sodium channel gene in permethrin-resistant *Aedes aegypti* and the distribution of this mutation throughout Thailand. *Trop Med Int Health*, *16*, 501–509. doi:10.1111/j.1365-3156. 2011.02725.x.

Yashutomi, K. & Takahasi, M. (1989). Insecticidal resistance of *Culex tritaeniorhynchus* in Chinen, Okinawa Prefecture, with special reference to the mechanism of pyrethroid-resistance. *Jpn J Sanit Zool*, *40*, 315–321 (in Japanese).

Zaim, M. & Jambulingam, P. (2007). *Global insecticide use for vector-borne disease control* (WHO/CDS/WHOPES/GCDPP/2002.2., World Health Organization). http:// apps.who.int/ iris/bitstream/ 10665/44220/1/ 9789241598781_eng.pdf?ua=1.

Biographical Sketch

Hitoshi Kawada

Affiliation: Division of Vector Ecology & Environment, Institute of Tropical Medicine, Nagasaki University, Nagasaki, Japan

Education: Doctor of Medicine (Nagasaki University), Doctor of Agriculture (Kyoto University)

Business Address: Sakamoto 1-12-4, Nagasaki, 852-8523 Japan

Research and Professional Experience:

1983 - 2002: Agricultural Chemicals Research Laboratory, Sumitomo Chemical Co., Ltd.

1. Developmental research of household and public health insecticides

2. Developmental research of insect growth regulator, pyriproxyfen
3. Developmental research of microencapsulated insecticides
4. Research of insect resistance to pyrethroids

2002 - Present: Institute of Tropical Medicine, Nagasaki University

1. Behaviological and ecological study on vector mosquitoes
2. Developmental research of new insecticide device for control of vector mosquitoes

Professional Appointments: Associate Professor

Honors:

1. Prize for Encouragement of Research of Japanese Society of Environmental Entomology and Zoology (1995)
2. Prize of Japan Society of Medical Entomology and Zoology (2014)

Publications Last 3 Years:

1. Kawada, H; Ohashi, K; Dida, GO; Sonye, G; Njenga, SM; Mwandawiro, C; Minakawa, N (2014) Preventive effect of permethrin-impregnated long-lasting insecticidal nets on the blood feeding of three major pyrethroid-resistant malaria vectors in western Kenya. *Parasit Vect*, *7*, 383.
2. Kawada, H; Ohashi, K; Dida, GO; Sonye, G; Njenga, SM; Mwandawiro, C; Minakawa, N (2014) Insecticidal and repellent activities of pyrethroids to the three major pyrethroid-resistant malaria vectors in western Kenya. *Parasit Vect*, *7*, 208.
3. Kawada, H (2014) Trials for the possible new controlling techniques as countermeasures against insecticide resistant vector mosquitoes. *Med Entolol Zool*, *65*, 45-59 (in Japanese).
4. Kawada, H; Oo, SZM; Thaung, S, Kawashima, E; Maung, YNM; Thu, HM; Thant, KZ; Minakawa, N (2014) Co-occurrence of the point mutations in voltage-gated sodium channel in pyrethroid resistant *Aedes aegypti* populations in Myanmar. *PLoS Neg Trop Dis*, *8*, e3032.
5. Kawada, H; Dida, GO; Ohashi, K; Kawashima, E; Sonye, G; Njenga, SM; Mwandawiro, C; Minakawa, N (2014) A small-scale field trial of pyriproxyfen-impregnated bed nets against pyrethroid-resistant *Anopheles gambiae* s.s. in western Kenya. *PLoS One*, *9*, e111195.

6. Kawada, H (2016) The present condition of disease vector mosquitoes and difficulties in vector control. *Sen'i Gakkaishi*, *72*, 148-154 (in Japanese).
7. Kawada, H; Higa, Y; Futami, K; Muranami, Y; Kawashima, E; Osei, JHN; Dadzie, S; Sakyi, KY; de Souza, D; Appawu, M; Ohta, N; Suzuki, T; Minakawa, N (2016) Discovery of point mutations in the voltage-gated sodium channel from African *Aedes aegypti* populations: Potential phylogenetic reasons for gene introgression. *PLoS Negl Trop Dis* 10: e0004780.
8. Kawada, H (2016) Possible new controlling measures for the pyrethroid-resistant malaria vectors. *Acta Horticulturae*, *in press*.

In: DDT
Editor: Kathleen Sanders
ISBN: 978-1-53610-009-9

Chapter 3

PHOTOCATALYTIC SELECTIVE OXIDATION OF DDT (DICHLORODIPHENYLTRICHLOROETHANE) TO DICOFOL

R. Bakhtchadjian**, L. A. Manucharova and L. A. Tavadyan
Nalbandyan Institute of Chemical Physics,
National Academy of Sciences of the Republic of Armenia,
Yerevan, Republic of Armenia

ABSTRACT

DDT (dichlorodiphenyltrichloroethane), used as a pesticide, is a persistent pollutant and rests one of the main dangers to human health and to the environment. Being chemically stable in ordinary conditions, its oxidation or oxidative degradation, especially, in large amounts or in industrial scale, is problematic in certain aspects. In this work it has been presented the experimental results of the direct oxidation of DDT with dioxygen, cheapest oxidant agent, using as a heterogeneous photocatalyst the fully characterized metal organic complex dioxo-Mo(VI)-

* Corresponding author: R. Bakhtchadjian. Address: Institute of Chemical Physics, National Academy of Sciences of the Republic of Armenia, Paruyr Sevak st. 5/2, 0014 Yerevan, Republic of Armenia. Email: robakh@hotmail.com.

dichloro[4,4′-dicarboxylato-2,2′-bipyridine], anchored on the surface of a TiO_2 by chemical bonds. The oxidation of DDT was carried out as a two stage cyclic process, involving: (i) the direct interaction of Mo-complex/TiO_2 with DDT under UV-irradiation (λ = 253.7 nm), and, (ii) regeneration of the photocatalyst with dioxygen in the dark conditions. These two stages of the process were separated temporally. It has been shown that the dioxo-Mo complex, anchored on the TiO_2, plays the role of photocatalyst in selective oxidation of DDT to dicofol (2,2,2-trichloro-1,1-*bis*-(4-chlorophenyl)ethanol) in chosen "mild" conditions. About 35% of the initial DDT was oxidized and the yield of dicofol was 21 mol %, during in 5 consecutive experimental cycles (32 h). It has been shown also that the formation of products, other than dicofol, is intermediated mainly by the oxidative and partially reductive decomposition of dicofol. They are dechlorinated (28-30%) and non-dechlorinated or partially dechlorinated organic compounds (34-36%), as well as the products of the complete degradation to CO_2 H_2O and HCl (15-17%). On the basis of kinetic data and analysis of products it has been proposed the probable mechanism of the photocatalyzed oxidation of DDT in the mentioned conditions. The scheme of the reaction mechanism includes the primary reaction of the oxo-atom transfer from the Mo-complex to tertiary carbon-atom of DDT by the formation of dicofol. About twofold increase of the turnover number of the reaction (Δ[DDT]/[anchored complex]) in second and further catalytic cycles evidences that the re-oxidation with dioxygen of Mo(IV) metal centers, formed in the first stage of catalytic cycle, to Mo(VI) may occur by the formation of intermediate Mo(VI)-oxo-peroxo moieties of the anchored complex.

Keywords: DDT, photocatalytic oxidation, dioxo-Mo(VI) complex/TiO_2, dicofol

INTRODUCTION

Some Aspects of the Oxidation of DDT with Dioxygen

Due to the chemical stability in ordinary conditions, DDT (dichlorodiphenyltrichloroethane) is known as a persistent organic pollutant (POP) [1]. The use of it as a pesticide causes a number of problems in environment, strongly affecting to the human health [2]. DDT is not oxidized with air or dioxygen in ordinary conditions. Moreover, some strong oxidants, such as chromic oxide in glacial acetic acid and nitric acid, are unable to oxidize DDT [3]. It is "quite resistant" also to oxidation by potassium permanganate [4], persulfate, hypochloride, and some other oxidants [3, p. 105].

Presently, the use of DDT is banned or reduced in majority of the countries. However, the problems related to the currently existing amounts of DDT in many countries rest actual. In spite of the successful development of AOP (Advanced Oxidation Processes) [5] in the treatment of water contaminated with relatively small amounts of DDT, they may not be applied for a large scale chemical transformation by several reasons, some of which were mentioned in the review [6] (for example, the impossibility to recover the photocatalyst by easy means). In this regard, at the present time, the nearly only chemical way of the complete destruction ("neutralization") of the great amounts of DDT rests its incineration. It is using in large industrial scale at the temperatures 900-1400°C [7, 3, p. 106]. However, the incineration is economically disadvantageous process, and, in his turn, poses some other environmental problems [7, 8]. In chemical industry, the oxidation of DDT, directed to the production of dicofol (2,2,2-trichloro-1,1-*bis*-(4-chlorophenyl)ethanol), is performed as a multistep process, including: the elimination of HCl from DDT, giving DDE (dichlorodiphenyldichloroethylene); chlorination of DDE to tetramer (dichlorodiphenyltetrachloroethane); hydrolysis of tetramer to dicofol, using para-toluene sulfuric acid [9].

In the chemical literature, in our knowledge, at least, in the recent decades, there are not works dedicated to investigations of the direct (with air or dioxygen) and selective oxidation of DDT to target products. Nearly all catalytic or photocatalytic investigations, devoted to the oxidation of DDT, mainly, were directed to implement the non-selective degradation or oxidative destruction of DDT in the frames of the environmental problems. In this regard, the aim and scope of the present work is deferent from the traditionally existing approaches in this field of investigations. The main aim of this chapter is the demonstration of the feasibility of the selective photocatalytic oxidation of DDT directly with dioxygen, in mild conditions.

Application of Some Transition Metal-Organic Complexes in Photocatalytic Oxidation of the Halogenated Derivatives of Aryl Hydrocarbons

DDT is a halogenated derivative of the aryl hydrocarbon. The feasibility of the photocatalytic oxidation of some halogenated derivatives of the aryl hydrocarbons with dioxygen, in mild conditions, was shown on the examples of the oxidative destruction of 1-chloro-4-ethylbenzene and 2,2-*bis*-(p-chloropheny)methane, by the application of the transition metal-organic complex, particularly, dioxo-Mo(VI) complex anchored on the surface of a TiO_2, under UV-irradiation [10-12]. How it occurs?

In general, the terminal oxo, dioxo, oxo-peroxo and peroxo complexes of some transition metals (Mo, W, V, Mn, Fe, Cr, etc.), having organic ligands, exhibit catalytic activity in the oxidation reactions of certain classes of the organic compounds (alkanes, alkenes, alcohols, Photocatalytic Selective Oxidation of DDT(dichlorodiphenyltrichloroethane) to Dicofol 3 aldehydes, etc.) in the presence of dioxygen [13-17]. Particularly, Mo-oxo complexes, bearing specific organic ligands, possess a capacity of the oxidation of aryl alkanes, phosphines, olefins and alcohols [18-21]. Analogously, some enzymes, containing similar transition metal-oxo- or dioxo moieties, also exhibit activity in the oxidation reactions of organic compounds with dioxygen [21, 22]. Apparently, there are some analogies in the oxidation mechanisms of these processes. Experimentally, it has been established that, in majority of cases, the primary chemical process begins by the oxo-atom transfer from these complexes (or cofactors of enzymes) to substrates [20, 21]. Oxidizing agents, such as O_2, H_2O_2, O_3, and others, in certain conditions, are able to re-oxidize the reduced metal center. As a result, in their presence, the processes become catalytic.

The catalytic activity increases, when these metal organic complexes are supported on the solid substances or anchored on the solid matrix with chemical bonds [23-25]. The metal oxides, such as TiO_2 and SiO_2, are more used support materials for grafted or anchored organometallic complexes in successful catalysis of the different chemical transformations [21, 26, 27]. Fully characterized complex dioxo-Mo(VI)-dichloro[4,4′-dicarboxylato-2,2′-bipyridine], anchored on a TiO_2 (Figure 1, Complex 1) is one of the best examples of them [23, 24].

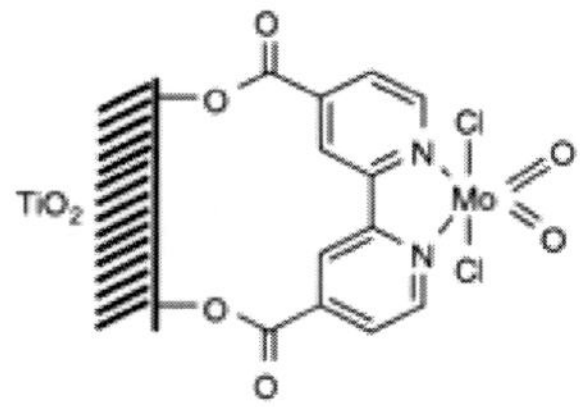

Figure 1. Complex 1. Dioxo-molybdenum(VI)-dichloro[4,4′-dicarboxylato-2,2′-bipyridine], anchored on TiO_2.

One of the characteristic properties of the great importance of Complex 1 and other analogous materials is their photo sensitivity under visible light or UV-irradiation. Therefore, the advantages of their use in photocatalytic processes is expected. Seemingly, the semiconductor support (TiO_2 in Complex 1, in the given case) may enhance the electronic flux to metallic centers of complex, under UV-irradiation or visible light [21, 23, 24]. On the other hand, the dioxo-Mo moieties in Complex 1 are isolated via an organic ligand, covalently bonded on the surface, and, the formation of Mo(V) species or its oligomers during the reaction is prevented [29].

The catalytic activity of Complex 1 has been tested in the reactions of oxidation of trimethylphosphine, ethylbenzene, tetraline, diphenylmethane [23-25, 28] with dioxygen, under UV-irradiation. The main products of these reactions were corresponding alcohols and ketones, the consequent partial decomposition of which under reaction conditions results the formation of other oxygenated products, including the products of the complete oxidation, CO_2 and water. It has been shown that for derivatives of the aromatic hydrocarbons the primary chemical act is the oxo-atom transfer from Complex 1 to the benzylic (or bibenzylic) carbon atom. According to the experimental results, dioxygen or other oxidant agents are able to oxidize the reduced form of metallic centers of the anchored complex ($Mo^{IV} \rightarrow Mo^{VI}$), even, in the dark conditions.

Above represented experimental results permit to assume the possibility of the direct oxidation of DDT with dioxygen by the application of the transition metal organic complexes of the mentioned types, under UV-irradiation, in quite mild conditions. In this regard, in the present work it has been investigated the photocatalytic oxidation of DDT with dioxygen, over dioxo-molybdenum(VI)-dichloro[4,4′-dicarboxylato-2,2′-bipyridine], anchored on TiO_2 as a two stage redox catalytic process. Experimentally, the reaction was carried out in the conditions of the temporal separation of the oxidation and reduction stages of the photocatalyst. The temporal separation of stages of the complex processes, widely used in the redox type heterogeneous-catalytic reactions, may be represented by the following simple scheme [30]

Substrate + Pseudo-catalyst (oxidant) → Oxygenated substrate + Pseudo-catalyst(reduced form) (I)

Pseudo-catalyst (reduced form) + O_2 → Pseudo-catalyst (oxidant) (II)

The first of these two consecutive stages is an oxidation reaction of any substrate by oxidant substance that may be named also pseudo-catalyst (metal oxides, complexes containing metal ions of variable valence with oxo or peroxo atoms, other oxidizer substances). The second stage, temporally separated from the primary one, is re-oxidation (regeneration) of the reduced form of the substance, used in a primary reaction, by oxidant agents such as dioxygen, hydrogen peroxide, etc. Two above reactions (I) and (II) compose a catalytic cycle. In the case of the oxidation of DDT with dioxygen, in our experiments, the first stage was a heterogeneous photochemical process between solid state metal organic Complex 1 (pseudo-catalyst, oxidant) and a substrate (DDT), under UV-irradiation, resulting to oxygenated product (mainly, dicofol) and a reduced form of the complex. Stopping the first stage, in certain moment, the reduced form of Complex 1 may be re-oxidized under dioxygen, in the second stage (in the given case, the second stage occurs without UV-irradiation). The temporal separation of the stages in redox catalysis or photocatalysis permits not only to control the process of the selective oxidation, but also to verify the agreement between the proposed reaction mechanism and the process.

The investigation of the detailed mechanism of above represented photocatalytic reactions have an importance not only for the synthetic chemistry, including industrial processes, but also for the modeling and understanding of enzymatic reactions, which are extensively spread in nature. The perspectives of the development in this field are related with the rational design and synthesis of new complexes, by sterically appropriate and specific organic ligands, anchored onto solid matrix.

EXPERIMENTAL

Investigation of the Photocatalytic Selective Oxidation of DDT to Dicofol with Dioxygen by the Application of the Dioxo-Molybdenum (VI) Complex, Anchored on TiO_2

Experimental Setup and Reactants

For the investigation of the oxidation of DDT with dioxygen by the participation of dioxo-Mo(VI)-dichloro[4,4′-dicarboxylato-2,2′-bipyridine], anchored on the surface of a TiO_2, a specially designed experimental setup, consisting of the photochemical glass reactor (quartz, 30 mL) with a magnetic stirrer and the optical junction for UV-irradiation (λ = 253.7 nm), high Photocatalytic Selective Oxidation of DDT(dichlorodiphenyltrichloroethane) to Dicofol 5 pressure Hg-lamp (DRT-230, Russia), cooling with distilled water, was used. It was described in [31].

DDT (1,1,1-trichloro-2,2-*bis*(4-chlorophenyl)ethane), dicofol (2,2,2-trichloro-1,1-*bis*-(4-chlorophenyl)ethanol), DDE (1,1-*bis*-(4-chlorophenyl)-2,2-dichloroethene), and nearly all other chemicals (Sigma-Aldrich) were used as received. Acetonitrile (Merck, Germany) was distillated by an ordinary procedure, in argon atmosphere and was kept with thermally activated molecular sieves.

Titanium dioxide (Degussa, P25, 80% anatase and 20% rutile) with a specific area about 50 m^2/g and particle size about 20 nm, was used for synthesis after the dehydration at 60°C and at 5×10^{-7} bar, during more than 48h.

The method of the synthesis of dioxo-molybdenum(VI)-dichloro-[4,4’-dicarboxylato-2,2’-bipyridine] complex, anchored on the TiO_2 surface, was described in [23]. The synthetic route of this complex was simplified using a nearly one-pot process: the transesterification of trimethylsilylated titania with the carboxylic ligand, giving trimethylsilanol (eliminated as hexamethyldisiloxane and water), then, the complexation was done by treating the anchored complex with a tetrahydrofuran solution, containing the calculated amount of MoO_2Cl_2. The latter was described in [25, 10, 12]. The characterization of the synthesized complex mainly was performed by ^{13}C CAP NMR method [25, 31]. The amounts of the anchored complex on the TiO_2 were determined by TGA analysis, using TG-2950 Hi Res. It was about $1,6 \times 10^{-4}$ mol per gram.

Procedures

Reaction mixture was a suspension of DDT (5,73 mg = 1.6×10^{-5} mol) in 10 mL acetonitrile, containing 10 mg of the anchored Complex 1. Each experimental cycle was consisted of two consecutive periods: in the first period, the reaction mixture was exposed under UV-irradiation in stirring conditions, in argon atmosphere and in the absence of dioxygen, during about 6-7 h, then, in the second period, the reaction was carried out without UV-irradiation and in the presence of dioxygen (2-2.5 h). Before the beginning of a new experimental cycle, dioxygen was thoroughly removed from the reaction mixture during 1.5-2 h, using inert gases, argon or helium. All experiments were carried out at the room temperature. Accordingly, the two periods of each experimental cycle correspond to the oxidation (I) and reduction (II) stages of the redox catalytic cycle, mentioned above.

Identification and analysis of the products were carried out by gas-liquid chromatography with electron capture detector, using column with SE-30 sorbent, the control analysis were carried out by chromatography-mass-spectrometer DANI Master DC TOF, using capillary column (30 m), coated by DB-ALC2. Analysis of Cl ions, were performed by chemical methods (argentometric way by the Mohr method).

DISCUSSION

First of all, note that the overall process may be considered as either heterogeneous or heterogeneous-homogeneous photocatalytic reaction, as Complex 1 is stable and not soluble in ordinary organic or inorganic solvents, particularly, in acetonitrile.

Experimental results of the consumption of DDT in 5 consecutive cycles are presented in Figure 2. The horizontal sections on the curve (a) correspond to the second periods ("dark" reaction), the aim of which is re-oxidation of Mo(IV), formed in the first period, into Mo(VI) with dioxygen. The curve (b) corresponds to the photochemical oxidation of DDT on the surface of "pure" TiO_2 (without anchored complex on TiO_2), in the presence of dioxygen. Note, the amount of the "pure" TiO_2 in reaction mixture was twice that in experiments with Complex 1. The simple comparison of two curves indicates a significant increase of the consumption of DDT in the presence of the anchored complex.

In reaction conditions, the turnover number, a molar ratio Δ[DDT]/[anchored complex], at the last experimental cycle is about 3.7 (5.8

times more than in the first cycle) (Table 1). In the course of 32 h reaction, the consumption of DDT is approximately 35%. In same conditions, but in the absence of the anchored complex, the photooxidation of DDT (even, in 2 times more quantities of "pure" TiO_2), is relatively slow process (Figure 2. b) and the consumption of DDT for the same time period of the reaction is nearly 13%. The above facts indicate the catalytic character of the overall process in the presence of the anchored Complex 1.

The reaction products include large spectra of the partially chlorinated and non-chlorinated organic compounds (C_1-C_{14}), as well as the products of the complete destruction, such as CO_2, Photocatalytic Selective Oxidation of DDT(dichlorodiphenyltrichloroethane) to Dicofol 7 H_2O, HCl. Among them, taking into account the results of the chromatographic analyses and overall amount of the analyzed Cl^- ions, the sum of a dozen products of the complete dechlorination was estimated as about 53-57%. Correspondingly, other products (43-47%) are non-dechlorinated and partially dechlorinated compounds.

The main product of the reaction is dicofol (21 mol %), playing also the role of intermediate for a number of other chlorinated and non-chlorinated products. Note, under the same conditions, but in the absence of the anchored complex (on the "pure" TiO_2), the formation of dicofol was very slow, and the reaction gives a more complex mixture of other oxidative/reductive decomposition products, including the products of the complete degradation in small amounts.

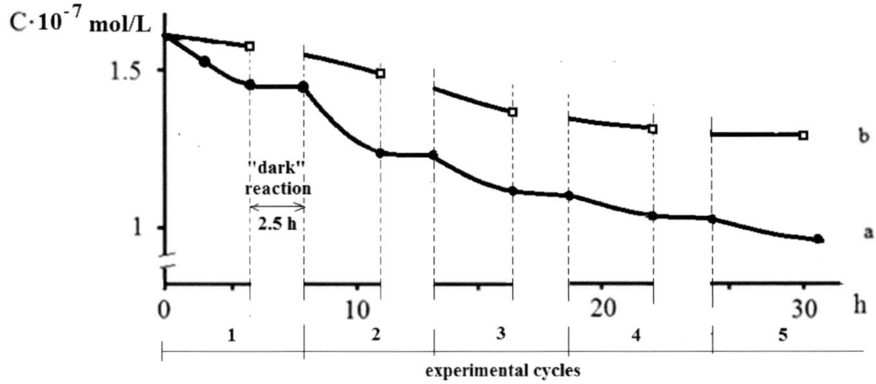

Figure 2. Time profiles of the consumption of DDT (C × 10^{-7} mol/L), under UV-irradiation: (a) - in the presence of the anchored Complex (1); (b) - in the presence of the "pure" TiO_2 (in amounts 2 times more than Complex 1) with O_2.

Table 1. Experimental data of the dependence of the turnover number Δ[DDT]/[anchored complex] and yield of dicofol on the number of experimental cycle (the reaction time (h) is given in parentheses)

Number of the experimental cycle	1	2	3	4	5
Turnover number	0.63 (5)	2.13 (11)	2.58 (17)	3.38 (23)	3.68 (32)
Yield of dicofol (mole %)	9.5	13.5	15.4	18.2	21

The primary reaction of the formation of dicofol, the main product of the reaction, may bepresented as follows:

$$(ClC_6H_4)_2C(H)CCl_3 + O{=}Mo^{VI}(O)(Cl_2)[L]/TiO_2 \rightarrow (ClC_6H_4)_2C(OH)CCl_3 + + Mo^{IV}(O)(Cl_2)[L]/TiO_2 \quad (1)$$

where L is 4,4′-dicarboxylato-2,2′-bipyridine.

As an indirect evidence of the formation of O = Mo(IV)-moieties in the reaction (1) may be served the spectroscopic data obtained in [28]. The same Complex 1, anchored on a mesoporous-TiO_2, was tested as a photocatalyst in the oxidation of diphenylmethane, under UV-irradiation and in analogous conditions. According to the authors, the DRIFTS (Diffuse reflectance infrared Fourier transform spectroscopy) spectrum of the post-reaction complex has a single, strong sharp band (920 cm^{-1}), characteristic for O = Mo (IV), while the initial material has 2 bands, corresponding to $O_2Mo(VI)$.

The yields of dicofol (Table 1) have an increasing tendency, correlating with the number of experimental cycles and turnover number Δ[DDT]/[anchored complex]. As it has been mentioned in our previous works [10, 12], the re-oxidation of Mo (IV) metallic centers of the anchored complex to Mo(VI) with dioxygen, in the second periods of the experimental cycles, probably, may occur via the formation of oxo-peroxo-Mo(VI) moieties:

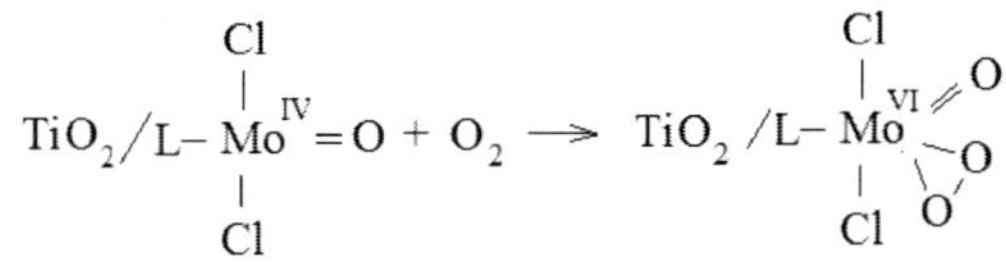

(2)

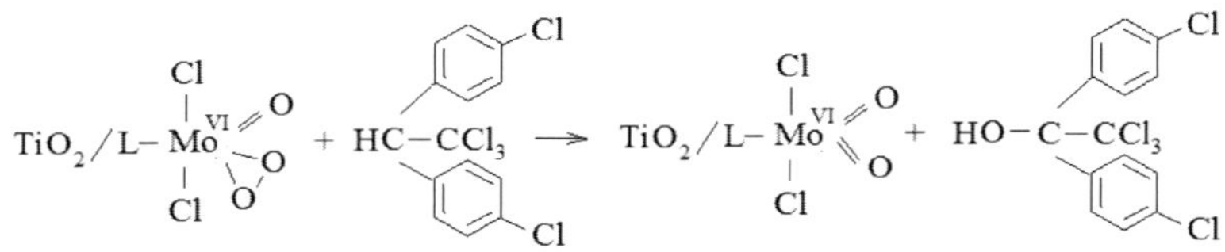

(3)

According to above scheme (reactions (2) and (3)), the amounts of the oxygen atoms, which are able to oxidize DDT (or other compounds present in the reaction mixture), will be doubled in the second and further experimental cycles in comparison with the initial amounts. This may be a cause of the fact that the turnover number in the second cycle increases more than two times, after the re-oxidation in the second period of the first cycle, becoming 2.13 from 0.63 (Table 1), while it could be no more than 2, or no more than 1 for only one cycle, taken separately, in consideration that the reaction (1) is a predominant way of the consumption of DDT. In other words, the consumption of DDT is in quantities more than initial stoichiometric quantities by the reaction (1), when one continue the reaction under UV-irradiation, after the "dark" reaction, in the absence of dioxygen. It is noteworthy that the yield of dicofole, also increases (about 1,42 times) with the increase of the consumed amount of DDT at the nearly same reaction time interval in the first period of the second cycle. In further experimental cycles, apparently, the different reaction channels of oxidation and decomposition become more significant, than the reactions (1)-(3).

As an indirect evidence of the reactions (1)-(3) may be served numerous examples of the analogous oxidation processes, of the chlorinated (1-chloro-4-ethylbenzene, 4,4'-dichlorodiphenylmethane) [10-12] and non-chlorinated hydrocarbons (ethylbezene, tetralin, diphenylmethane) [24, 25, 28], in same or similar experimental conditions. In all mentioned cases, the oxo-atom transfer from the anchored complex to the benzylic or bibenzylic carbon atom is the main primary reaction, resulting to the formation of the corresponding alcoholic compound.

The second main product of the reaction is DDE (1,1-*bis*-(4-chlorophenyl)-2,2-dichloroethene) the maximal yield of which is about 9 mol

%. DDE may be formed, at least, by several different channels. One of them may be the known dehypochlorination of dicofol under UV-irradiation or sunlight [32]

$$(ClC_6H_4)_2C(OH)CCl_3 \rightarrow (ClC_6H_4)_2C{=}CCl_2 + HOCl \quad (4)$$

$$2HOCl \rightarrow O_2 + 2H^+ + 2Cl^- \quad (5)$$

Other way of the formation of DDE may be the reaction of dicofol with OH radicals, photo-generated on the TiO_2 from the water molecules. The mechanism of the photo-generation of OH radicals from water on the TiO_2 surfaces is well established in a great number of investigations in comparable chemical systems [33, 34]. According to the opinion of the authors [33], the degradation of dicofol, carried out under UV-irradiation, but in the absence of any previously adsorbed or chemisorbed compound on the surface of TiO_2, occurs via the formation of *bis*-(4-chlorophenyl)methanon and continues up to the final products of the degradation (CO_2, H_2O, Cl^-) by free radical reactions. The hydroxyl or hydroperoxy radicals in the reaction scheme, given in [33], were considered as the leading active species. In any case, it is evident, a decomposition of the carbonic chain of dicafol. One of the two expected products of the further oxidative decomposition of dicofol is dichlorobenzophenone $(ClC_6H_4)_2C{=}O$, identified in our experiments. The formation of the second product $ClC_6H_4C(O)CCl_3$ (2,2,2-trichloro-1(4-chlorophenyl)ethanone, non-observed in our experiments) is less probable by the reasons mentioned in [33]. The atom of carbon bonded with 3 chlorine atoms in dicofol, bears the highest electronic density, therefore, the cleavage of C-C bond between this carbon and carbon of $C(OH)(PhCl)_2$ by the attack of OH (or HO_2) radicals is more expected. Correspondingly, the products from dicofol will be $Cl(C_6H_4)_2C{=}O$, H_2O and radical CCl_3, finally, giving also CO_2, H^+ and Cl^- in the presence of dioxygen.

In preparative aspect, we paid attention to the waterlessness of initial reactants in the gas, liquid or solid states, practically assuring the absence of water in notable quantities. However, water forms in the course of the reaction as a final product. The appearance of water in the reaction media under UV-irradiation may initiate photochemical radical reactions with the formation of a great number of destruction products from dicofol, as well as DDE, on the surface of TiO_2 or in the liquid phase.

All other products of the oxidative decomposition in our experiments, in general, are same that were observed in the photochemical decomposition of dicofol, investigated in [33]. However, the composition of products in the presence of the anchored complex on a TiO_2 has Photocatalytic Selective Oxidation of DDT(dichlorodiphenyltrichloroethane) to Dicofol 9 some obvious differences. After 32 h reaction under UV-irradiation, the reaction mixture contains also products of the reductive decomposition, such as DDD (1,1-dichloro-2,2-*bis*-(p-chlorophenyl)ethane), DDM (2,2-*bis*-(p-chloropheny) methane, diphenylmethane), and others, summarily about 13-14%. Actually, we are not able to suggest any substantiated explanation to this observation. Note only, the presence of some similarity in the composition of the reaction products between the enzymatic degradation of DDT and the results of the present work. For example: in biological degradation of DDT by so called White Rot Fungus Phanerochaete Chrysosporium, the formation not only products of the oxidative (2,2-dichloro-1,1-bis(4-chlorophenyl)ethanol (known as FW-152), and 4,4'-dichlorobenzophenone (DBP)), but also of reductive degradation (DDD, DDM, and others) were identified [35].

Summarizing the above data and mentioned observations, it may be assumed a complex mechanism of the oxidation of DDT with O_2 in the presence of Complex 1, under UV-irradiation. Apparently, the mechanism include a combination of the primary transfer of oxo- or peroxo-atoms of Mo(VI) to the substrate, accompanied with a number of further decomposition and oxidation reactions.

CONCLUSION

The obtained experimental results show the possibility of the direct photocatalytic oxidation of DDT with dioxygen in quite mild conditions, and, it may be considered as a process, corresponding to the some important requirements of the "green" chemistry. This feasibility opens an attractive way for the elaboration of an alternative method of the oxidation of DDT in mild conditions, using other anchored complexes, containing the transition metal oxo-atoms, on a TiO_2, under UV-irradiation or sunlight. In this regard, the application of dioxo-Mo(VI) complex/TiO_2, as a photocatalyst, exhibiting activity and selectivity in the oxidation of DDT to dicofol, in the presence of dioxygen, is an example of the successful catalysis. One of the important evidences of this is continuous increase of the turnover number, which is more than unity (Δ[DDT]/[anchored complex]>1), beginning from the second experimental cycle.

The mentioned reaction also is a good example of the heterogenization of the homogeneous catalysts, working in photochemical conditions. Moreover, the anchoring of the given transition metal complexes onto the photosensitive support materials, as TiO_2, remarkably improves the catalytic activity of Mo-dioxo complex toward a number of organic compounds, including, even, quite "resistant" substances to oxidation, as DDT. Apparently, the selective oxidation of DDT to dicofol occurs via oxo-atom transfer from dioxo-Mo(VI) moieties to the bibenzylic carbon-atom, as a primary reaction. However, the further consecutive oxidation and decomposition, even, reduction reactions of dicofol in photochemical conditions, become cause of a large spectra of other chlorinated and non-chlorinated products, including products of the complete oxidation of DDT.

The described reaction is a typically redox photocatalytic process, the oxidation and reduction stages of which may be separated temporally in experimental investigations. From the point of view of the understanding of the reaction mechanism, the temporal separation of stages permits directly to investigate both the reduction of $O_2Mo(VI)$ by substrate and re-oxidation of O = Mo(IV) by dioxygen, correspondingly. The formation of peroxo moieties, in this case, seemingly, is more probable pathway of re-oxidation. However, in the present state, it is a working hypothesis, although there are indirect experimental evidences about their formation.

Some analogies in the composition of the reaction products between the enzymatic oxidative decomposition and the results of the direct oxidation by dioxo-Mo(VI)/TiO_2 complex of DDT were observed. They also need further investigations in the aims of the explanations of these and other analogies, related to the selective reactions of the oxo-atom transfer of transition metal organic complexes.

From the practical point of view, the investigation of this reactions open a new opportunity for highly selective oxidation of DDT to desired products in industrial scale, fully corresponding to the critters of the "green" chemistry.

REFERENCES

[1] Stockholm Convention on Persistent Organic Pollutants (POPs), as amended in 2009. *Text and Annexes*, UN, pp. 1-64.

[2] World Health Organization, *DDT in Indoor Residual Spraying: Human Health Aspects.* (2011). pp. 1-309. Website: www.who.int/ipcs/publications/ehc/ehc241.pdf.

[3] Lawless, E. W.; Ferguson, T. L.; Meiners, A. F. *Guidelines for the Disposal of Small Quantities of Unused Pesticides, Part A. Pesticides, Pesticides Chemistry, and Pesticide Disposal.* Section 7, Review of the chemistry of pesticide disposal (pp. 53-134), DDT (pp. 104-105). Environmental Protection Agency, National Environmental Research Centre, Office of Research and Development, Cincinnati, OH, 1975, pp. 1-335.

[4] DDT Stocks, Collection, Transportation and Disposal: Environmental Impact Statement (Prepared by Louise Berger and Associates). *History of DDT* (p. 19-91), US Department of Stocks of DDT, Battle Creek, MI, 1981, p. 85.

[5] Parsons, S. A.; Byrne, A. Water Treatment Applications (section 13. pp 329-346), *Pesticides*, p. 336. In: Parsons S. (ed.). Advances Oxidation Processes in Water and Wastewater Treatment. IWA Publishing, London, 2004, pp. 1-356.

[6] Das R. Application Photocatalysis for Treatment of Industrial Waste Water - A Short Review. *Open Access Library Journal*, 1, 2014, p. 713. Website: http://dx.doi.org/10.4236/oalib.1100713.

[7] Munnecke, D. M., (1979). Chemical, physical, biological methods of the disposal and detoxification of pesticides. *Residue Reviews, Residues of Pesticides and Other Contaminants in the Total Environment* 70, 1-26, p. 10.

[8] Adeola, F. Hazardous Wastes, Industrial Disasters, and Environmental Health Risks. *Local and Global Struggles*, Palgrave Macmillan, NY, 2011, pp. 1-235, p. 45.

[9] Srikumar, K. *Chemical Process Technology and Simulation*, PHI Leaming Private Limited, Dehli, 2013, pp. 1-352, p. 127.

[10] Bakhtchadjyan, R.; Tsarukyan, S.; Barrault, J.; Matinez, F.; Tavadyan, L.; Castellanos, N., (2011). Application of a dioxo-molybdenum(VI) complex anchored on TiO_2 for the photochemical oxidative decomposition of 1-chloro-4-ethylbenzene under O_2. *Transition Metal Chemistry* 36, 8, 897-900.

[11] Bakhtchadjian. R. H.; Tsarukyan, S. V.; Manucharova, L. A.; Tavadyan, L. A.; Barrault J., (2012). Dioxo-Mo(VI) - Complexes Anchored on Titania as Catalysts in Photo-Oxidative Decomposition of Some Aromatic Organochlorine Compounds by Dioxygen. *IX International Conference "Mechanisms of Catalytic Reactions."* October 22-25, 2012. St. Petersburg, Abstracts (CD, PP-V-3), p. 70.

[12] Bakhtchadjyan, R. A.; Tsarukyan, S. V.; Manucharova, L. A.; Tavadyan, L. A.; Barrault, J.; Photocatalytic Selective Oxidation of DDT(dichlorodiphenyltrichloroethane) to Dicofol 11. Matinez, F. O., (2013). Photochemical decomposition of 1 chloro-4-ethylbenzene by participation of O_2 and a dioxomolybdenum(VI) complex anchored on the TiO_2 surface. *Kinetics and Catalysis*. 54, 1, 34-39.

[13] Meunier B. (ed.) Metal-Oxo and Metal-Peroxo Species in Catalytic Oxidations. *Structure and Bonding.* Springer-Verlag, Berlin, Heidelberg, 2000. Vol. 97, pp. 1-329.

[14] Huber, S.; Cokoja, M.; Kühn, F. E., (2014). Historical landmarks of the application of molecular transition metal catalysts for olefin epoxidation. *Journal of Organometallic Chemistry* 751, 25-32.

[15] Chiusoli, G P.; Maitlis, P. M. (eds.). *Metal-catalysis in Industrial Organic Processes.* RCS-Publishing, Cambridge, 2006, pp. 1-257.

[16] Simandi, L. I. *Catalytic Activation of Dioxygen by Metal Complexes*. Kluwer Academic Publishers, Dordrecht, Netherlands, 1992, pp. 1-401.

[17] Waidmann, C. R.; DiPasquale, A. G.; Mayer, J. M., (2010). Synthesis and reactivity of oxo-peroxo-vanadium(V) bipyridine compounds. *Inorg. Chem.* 49, 5, 2383-2391.

[18] Arzoumanian, H., (1998). Molibdenum -oxo chemistryin various aspects of oxygen atom transfer. *Coord. Chem. Rev.* 178-180, 1, 191-202.

[19] Paez-Mozo, E. A., (2006). Efecto de la sensibilización en oxidación foto-catalítica *Rev. Acad. Colomb. Cienc.* 30, 116, 407-413.

[20] Arzoumanian, H., (2011). Molibdenum-oxo and peroxo complexes in oxygen atom transfer processes with O_2 as a primary oxidant. *Current Inorganic Chemistry*. 1, 2, 140-145.

[21] Arzoumanian, H.; Bakhtchadjian, R., (2012). Oxo-Atom transfer reactions of transition metal complexes in catalytic oxidation with O_2 on the light of some recent results in molybdenum-oxo chemistry. *Chemical Journal of Armenia* 65, 2, 168-188.

[22] Visser S. P., Kumar D. *Iron-Containing Enzymes: Versatile Catalysts of Hydroxylation Reactions in Versatile Catalysts of Hydroxylation Reactions in Nature*. RSC of Chemistry, Cambridge, 2011, pp. 1-462.

[23] Paez, C. A.; Castellanos, N. J.; Martınez, F. O.; Ziarelli, F.; Agrifoglio, G.; Paez-Mozo, E. A.; Arzoumanian, H., (2008). Oxygen atom transfer photocatalyzed by molybdenum(VI) dioxodibromo-(4,4′-dicarboxylate-2,2′-bipyridine) anchored on TiO_2. *Catal. Today* 133-135, 619-624.

[24] Paez, C. A.; Lozada, O.; Castellanos, N. J.; Martınez, F. O.; Ziarelli, F.; Agrifoglio, G.; Paez-Mozo, E. A.; Arzoumanian, H., (2009). Arylalkane

photo-oxidation under visible light and O_2 catalyzed by molybdenum(VI)dioxo-dibromo (4,4′-dicarboxylato-2,2′-bipyridine) anchored on TiO_2. *J. Mol. Catal. A: Chem.* 299, 1-2, 53-59.

[25] Arzoumanian, H.; Castellanos, N. J.; Martınez, F. O.; Paez-Mozo, E. A.; Ziarelli, F., (2010). Silicon-assisted direct covalent grafting on metal oxide surfaces: Synthesis and Characterization of carboxylate N,N′-ligands on TiO_2. *Europ. J. Inorg. Chem.* 11, 1633-1641.

[26] García, N.; Benito, E.; Guzman, J.; Tiemblo, P., (2007). Use of p-toluenesulfonic acid for the controlled grafting of alkoxysilanes onto silanol containing surfaces: preparation of tunable hydrophilic, hydrophobic, and super-hydrophobic silica. *J. Am. Chem. Soc.* 129, 16, 5052-5060.

[27] Shylesh, S.; Singh, A. P., (2004). Synthesis, characterization, and catalytic activity of vanadium-incorporated, grafted, and immobilized mesoporous MCM-41 in the oxidation of aromatics. *J. Catalysis* 228, 2, 333-346.

[28] Castellanos, N. J.; Martınez, F.; Lynen, F.; Biswas, S.; Van Der Voort, P.; Arzoumanian, H, (2013). Dioxygen activation in photooxidation of diphenylmethane by a dioxo-molybdenum(VI) complex anchored covalently onto mesoporous titania. *Trans. Met. Chem.* 38, 119-127.

[29] Arzoumanian, H.; Bakhtchadjian, R.; Atencio, R.; Briceno, A.; Agrifolio, G., (2006). Characterizationof a reduced molybdenum-oxo compound derived from an oxo-transfer process under stoichiometric conditions *Journal of Molecular Catalysis A: Chemical* 260, 1-2, 197-206.

[30] Buzzoni, R.; Ricci, M.; Rossini, S.; Perego, C.; Seletive Oxidation of Eni. In: Duprez D. (ed.), Cavani F. (ed). *Handbook of Advanced Methods and Processes in Oxidation Catalysis from Laboratory to Industry*. Imperial College Press, London, 2014, pp. 1-1036, p. 364.

[31] Bakhtchadjyan, R. A.; Tsarukyan, S. V.; Manucharova, L. A.; Tavadyan, L. A.; Barrault, J.; Matinez, F. O., (2011). Photochemical oxidative decomposition of 1-chloro-4-ethylbenzene in the presence of dioxo-molybdenum(VI) complex anchored on the TiO_2. *Chemical Journal of Armenia* 64, 1, 9-15.

[32] Zepp, R. G.; Wolfe, N. L.; Azarraga, L. V.; Cox, R. H.; Pape, C. W., (1977). Photochemical transformation of the DDT and methoxychlor degradation products, DDE and DMDE, by sunlight. *Arch. Environ. Contam. Toxicol.* 6, 2-3, 305-314.

[33] Yu, B. B.; Zeng, J. B.; Gong, L. F.; Yang, X. Q.; Zhang, L. M.; Chen, X., (2008). Photocatalytic degradation investigation of dicofol. *Chines Science Bulletin* 53, 1, 27-32.

[34] Zhang, J.; Nosaka, Y. J., (2014). Mechanism of the OH radical generation in photocatalysis with TiO_2 of different crystalline types. *Phys. Chem. C.* 118, 20, 10824-10832.

[35] Bumpust, J. A.; Austt, S. D., (1987). Biodegradation of DDT [1,1,1-trichloro-2,2-bis(4-chlorophenyl)ethane] by the White Rot Fungus Phanerochaete Chrysosporium. *Applied and Environmental Microbiology* 53, 2001-2008.

Biographical Sketch

Robert Bakhtchadjian

Affiliation: Institute of Chemical Physics, National Academy of Sciences of the Republic of Armenia, Yerevan, Republic of Armenia; *Centre BA,* Marseilles, France

Education:

- Resolution of the National Commission of Equivalence (French Minister of National Education, Scientific Researches and New Technology Department, Montpellier University II) (Docteur ès Science) Montpellier (France) 2000
- Official Report of the Commission of Equivalence of ORSTOM (Paris, France); Docteur ès Science (en chimie), Paris, 1996
- Post-doctoral Fellow, Attestation (Centre National de la Recherche Scientific and Aix-Marseilles University, Marseilles, 1996-98
- Diploma of Doctor of Philosophy in chemistry (Scientific Grade "Kandidat nauk"), Highest Attestation Commission of the USSR, Moscow, 1986
- Certificate of the Academy of Sciences of Armenian SSR (specialty "Chemical Kinetics and Catalysis"), Yerevan, 1986
- Diploma (with Honors) of the Yerevan State University, Department of Chemistry, Chemist (MSc) degree, Yerevan, 1979

Research and Professional Experience

- Ph. Doctor (Chemistry-*Chemical kinetics and catalysis*), Moscow, 1986
- Docteur ès Sciences (*équivalence*), Paris, 1996; Montpellier, 2000
- Senior Scientific Researcher, Head of the International Scientific Group CNRS (France)/SCS (Armenia), Guest Professor. His scientific activity is linked with the International Research Cooperation programs between the scientific institutions of France and Armenia.
- Author or co-author of: 72 scientific publications (in peer-reviewed journals) in English, Russian, French, and Armenian, 3 Authorship Certificates, 1 monograph. Participation in 30 international scientific gatherings. A great number of popular-scientific articles in press.

Professional Appointments

- Nalbandyan Institute of Chemical Physics, National Academy of Sciences of the Republic of Armenia
- 2009-12 Head of the International Research Group of Scientific Co-operation CNRS (France)/SCS(Armenia),
- 2008-11 Senior Scientific Researcher (permanent)
- 2011 - Senior Scientific Researcher (permanent and freelance)
- European Regional Educational Academy, Armenia
- 2010 - Guest professor, course: "Principles of Environmental Physics and Phemistry"
- CNRS (UMR 6516 "Catalysis") Marseilles, in: National Higher School of Synthesis, Chemical Processes and Engineering of Aix-Marseilles (Aix-Marseilles University III)
- 1996-1998 Post-doctoral Fellow (fellowship of CNRS)
- 1997-1998 Engineer of Research
- 1999-2004 Researcher Chemist (determined contracts) CNRS (UMR 6516)
- Alpha Maintenance-*Ingénierie de maintenance – Chimie – Pétrochimie*" (SA, Martigues, Bouches-du-Rhône)
- 2004-2005 Deputy Director of the South Section
- Institute of Chemical Physics of the Academy of Sciences of the Armenian SSR, Laboratory of Chain and Radical Reactions, Yerevan, 1979-1995
- 1980-1986 Junior Scientific Researcher
- 1986-1989 Scientific Researcher

- 1989-1995 Senior Scientific Researcher

Awards

- Winner of the bilateral concours of Scientific Cooperation Project between France and the Republic of Armenia, 2009.
- Award of the Economic Commission of the *Provence-Metropole* (by Senator, Mayor of Marseille Jean Claude Godin, Marseille, France 1999)
- Distinguished Leadership Award, American Biographical Institute (US, 1999)
- First Prize for the best scientific article of the Council of Young Scientists, Yerevan, Armenia, 1982

Grants

- Grant of the International Sciences Foundation, US (1994)
- Grant of the Community of Commons, France (1997)
- Bilateral grant of scientific cooperation CNRS (France)/SCS(Republic of Armenia) 2009

Memberships

Professional Organization

- Member of the New York Academy of Sciences (US, NY, July 1995)
- Member of the International Academy of Sciences "Ararat" (Paris, 2003)
- Member of the American Association of Science and Technology (US, 2014)
- Member of the Editorial Boards of International scientific journals (2013-)

Non-Professional Organizations

- Founder and President of the "Armenian Biographical Center" (Marseilles, 2003)
- Member of the Editorial Council; American Biographical Institute (US, NJ, 2002)
- Elected Honorary Member of the Research Board of Advisors, American; Biographical Institute (US, 1998)

Biographical references: in more than 20 international biographical editions including:

- Who's Who in the World, 15-20 Marquis Ed., US, 1998-2002
- Outstanding People of 20th Century, IBC, Cambridge, UK 1999
- International Directory of Distinguish Leadership, ABI, US, 1998
- Who's who in Science and Technology, Marquis, US, 2004

Recent Publications

[1] Selective oxidation of DDT by dioxygen on the dioxo-Mo(VI) complex anchored on a TiO_2 under UV-irradiation (with Manucharova L.A., Tavadyan L.A.). *Catalysis Communications*, 2015, vol. 69, pp. 193-195.

[2] Photochemical decomposition of 1-chloro-4-ethylbenzene by participation of O_2 and a dioxomolybdenum(VI) complex anchored on the TiO_2 surface (with Tsarukyan S.V., Manucharova L.A., Tavadyan L.A. Barrault J., Matinez F.O.) *Kinetics and Catalysis*. 2013, vol. 54, 1, pp. 34-39.

[3] Oxo-atom Transfer Reactions of Transition Metal Complexes in Catalytic Oxidation with O_2 on the Light of Some Recent Results in Molybdenum-oxo Chemistry (a review), (with Arzoumanian H.) *Chemical Journal of Armenia*, vol. 65, 2, 2012, pp. 168-188.

[4] Application of a dioxo-molybdenum(VI) complex anchored on TiO_2 for the Photochemical Oxidative Decomposition of 1-chloro-4-ethylbenzene under O_2 (with Tsarukyan S., Barrault J., Matinez F., Tavadyan L., Castellanos N.). *Transition Metal Chemistry*. 2011, vol. 36, 8, pp. 897-900.

[5] Investigation of the catalytic activity of oxo-molibdenum catalyst anchored on TiO_2 in photooxidation reaction of aromatic hydrocarbons (with Manucharova L.A., Sargsyan T.A., Tavadyan L.A.): In: *Current Problems of the Chemical Physics*, Institute of Chemical Physics, NAS of Armenia, Yerevan, 2015, pp. 102-103.

[6] Photochemical Oxidative Decomposition of 1-chloro-4-ethylbenzene in the Presence of Dioxo-molybdenum(VI) Complex Anchored on the TiO_2 (with Tsarukyan S.V., Manucharova L.A., Tavadyan, L.A., Barrault J., Matinez F.O.). *Chemical Journal of Armenia*. 2011, vol. 64, 1, pp. 9-15.

[7] Some Examples of an Oxo-Atom Transfer Reactions of Molybdenum Complexes, Anchored on TiO_2 in Catalytic Photooxidation by O_2 (with Tsarukyan S.V., Manucharova L.A.). In: "Current Problems of Chemical Physics" (dedicated to the 50th anniversary of the Institute of Chemical Physics NAS RA), Armenia, Yerevan, 2012, pp. 142-143.

Robert Bakhtchadjian (in a number of the English publications: R.H. Bakhchadjyan), Ph. Doctor in chemistry (Moscow, 1986), *Docteur ès Sciences* (Paris, 1996 and Montpellier, 2000), is Senior Scientific Researcher (Institute of Chemical Physics), Head of the International Scientific Group CNRS (France)/SCS (Armenia), Guest Professor. His scientific activity is is related to the International Research Cooperation programs between the scientific institutions of France and Armenia.

His scientific interests are focused mainly on two areas of the chemistry: chemical kinetics and catalysis (chain and radical reactions, slow oxidation processes, heterogeneous reactions of free radicals in oxidation), and, chemistry of complex compounds (synthesis and characterization of transition-metal organic compounds, homogeneous and heterogeneous catalysis by complex compounds of transition metals). He is author or co-author of 72 scientific publications (in peer-reviewed journals) in English, Russian, French, and Armenian, 3 Authorship Certificates, dozens of popular-scientific articles, and 1 monograph. Scientific works were presented in more than 30 international scientific gatherings.

His memberships include professional scientific organizations: the New York Academy of Sciences (US, 1995), International Academy of Sciences "*Ararat*" (France, 2005), American Association of Science and Technology (2014), Editorial Boards of International scientific journals, as well as non-professional organizations (Armenian Biographical Center (founder and President, Marseilles, 2003; Research Board of Advisors (Elected Honorary Member) of the American Biographical Institute, 1998, etc).

He is recipient of a number of awards, prizes and grants, including: Award of the Economic Commission of the *Provence-Metropole* (by the Mayor of Marseille, Marseille, France, 1999), Winner of the bilateral concours for Scientific Cooperation Project between France and the Republic of Armenia (CNRS, 2009), Distinguished Leadership Award (ABI., 2009). First Prize for the best scientific article (Council of Young Scientists of Armenia, 1984), Grant of International Sciences Foundation (US, 1994), Grant of Community of Commons (France, 1998), Bilateral grants of scientific cooperation CNRS/SCS (2010).

INDEX

A

B

C

D

E

F

G

H

I

J

K

L

M

N

O

S

T